FOCUS ON

Grades K-4

ELEMENTARY

ASTRONOMY

3rd Edition

Rebecca W. Keller, PhD

REAL SCIENCE 4 Kids

Real Science-4-Kids

Illustrations: Janet Moneymaker

Focus On Elementary Astronomy Student Textbook—3rd Edition (softcover)
ISBN 978-1-941181-30-0

Published by Gravitas Publications Inc.
www.gravitaspublications.com
www.realscience4kids.com

GRAVITAS
PUBLICATIONS

Contents

CHAPTER 1   EXPLORING THE COSMOS   1

  1.1   Introduction   2
  1.2   Who Was the First Astronomer?   3
  1.3   Famous Early Astronomers   5
  1.4   Astronomers Today   8
  1.5   Summary   9
  1.6   Some Things to Think About   9

CHAPTER 2   ASTRONOMER'S TOOLBOX   10

  2.1   Introduction   11
  2.2   Brief History of a Basic Tool   11
  2.3   What Is a Telescope?   13
  2.4   Early Telescopes   14
  2.5   Advanced Telescopes   15
  2.6   Summary   16
  2.7   Some Things to Think About   16

CHAPTER 3   EARTH'S HOME IN SPACE   18

  3.1   Introduction   19
  3.2   Earth Is a Planet   20
  3.3   The Moon and Tides   23
  3.4   The Sun and Weather   24
  3.5   Eclipses   25
  3.6   Summary   25
  3.7   Some Things to Think About   26

**CHAPTER 4   EARTH'S NEIGHBORS—MOON
AND SUN**   27

  4.1   Introduction   28
  4.2   The Moon   28
  4.3   The Sun   30
  4.4   The Sun's Energy   32
  4.5   Summary   33
  4.6   Some Things to Think About   34

CHAPTER 5 EARTH'S NEIGHBORS—PLANETS   35

5.1 Introduction   36
5.2 Planets   36
5.3 Two Types of Planets   38
5.4 Where's Pluto?   41
5.5 Summary   42
5.6 Some Things to Think About   42

CHAPTER 6 OBSERVING CONSTELLATIONS   44

6.1 Introduction   45
6.2 Northern Hemisphere Constellations   47
6.3 Southern Hemisphere Constellations   49
6.4 Using Stars to Navigate   50
6.5 Summary   53
6.6 Some Things to Think About   53

CHAPTER 7 EARTH'S NEIGHBORHOOD   54

7.1 Introduction   55
7.2 Our Solar Neighborhood   55
7.3 Orbits   58
7.4 Why Is Earth Special?   61
7.5 Summary   62
7.6 Some Things to Think About   62

CHAPTER 8 BEYOND THE NEIGHBORHOOD   64

8.1 Introduction   65
8.2 Nearest Star   65
8.3 Brightest Star   67
8.4 Biggest Star   68
8.5 Stars With Planets   69
8.6 Summary   70
8.7 Some Things to Think About   70

CHAPTER 9 GALAXIES   71

9.1 Introduction   72
9.2 Galaxies Are Like Cities in Space   72
9.3 How Many Galaxies?   73
9.4 What Is a Galaxy Made Of?   74
9.5 Other Stuff About Galaxies   75
9.6 Summary   76
9.7 Some Things to Think About   76

CHAPTER 10 OUR GALAXY—THE MILKY WAY   77

10.1 Introduction   78
10.2 Our Galaxy   78
10.3 Where Are We?   80
10.4 Earth Moves   81
10.5 Summary   82
10.6 Some Things to Think About   82

CHAPTER 11 BEYOND OUR GALAXY   84

11.1 Introduction   85
11.2 More Spiral Galaxies   85
11.3 Other Types of Galaxies   86
11.4 The Local Group of Galaxies   88
11.5 Summary   89
11.6 Some Things to Think About   89

CHAPTER 12 OTHER STUFF IN SPACE   91

12.1 Introduction   92
12.2 Comets and Asteroids   93
12.3 Exploding Stars   94
12.4 Collapsed Stars   95
12.5 Nebulae   95
12.6 Summary   97
12.7 Some Things to Think About   97

GLOSSARY-INDEX   99

Chapter 1 Exploring the Cosmos

1.1 Introduction   **2**

1.2 Who Was the First Astronomer?   **3**

1.3 Famous Early Astronomers   **5**

1.4 Astronomers Today   **8**

1.5 Summary   **9**

**1.6 Some Things
     to Think About**   **9**

1.1 Introduction

Astronomy is the study of the cosmos. The term cosmos refers to the Earth and everything that extends beyond the Earth, including other planets, stars, nebulae, comets, asteroids, and even black holes.

Astronomy is a fascinating science with many new objects and areas of space waiting to be explored.

In this chapter we will learn about some well-known astronomers and their discoveries. We will also find out what skills are used by astronomers to study objects in space.

1.2 Who Was the First Astronomer?

Because astronomers can't fly to faraway planets or ride asteroids, astronomers use various tools and techniques to find out more about the objects in the cosmos. However, before the use of modern tools, people could learn a great deal about the cosmos by studying the night sky.

It's hard to say who was the first astronomer. Many early people studied the planets and stars, and even without modern tools they discovered a great deal about the cosmos.

Early Egyptian, Babylonian, and Mayan people observed the sky in great detail. Noting when the Moon was full

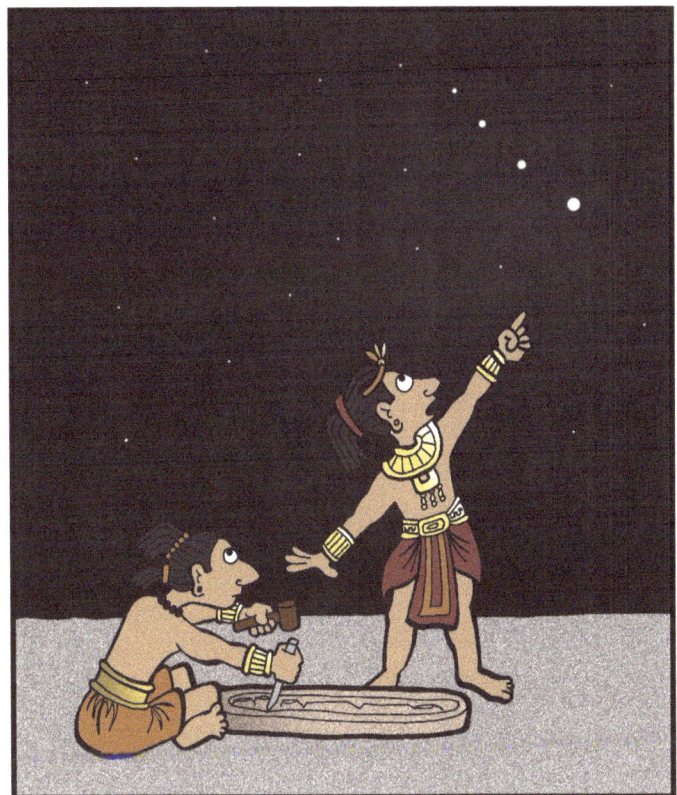

or when the Sun sank lower on the horizon, early observers were able to learn about how the planets and the Moon moved. From their observations they produced calendars and were able to predict eclipses.

One of the questions early astronomers asked was, "Does the Earth move around the Sun, or does the Sun move around the Earth?" In other words, do we live in a "Sun-centered" cosmos or an "Earth-centered" cosmos? To early astronomers it appeared from simple observation that we live in an Earth-centered cosmos. When the Sun rises and sets each day, it has the appearance of moving around the Earth. However, as we will see, sometimes how things move isn't always easy to figure out.

One of the very first astronomers to propose that the Earth moves around the Sun was Aristarchus of Samos. Aristarchus was a Greek astronomer and mathematician who lived from 310-230 BCE. He studied the planets and said that the Earth has two different movements. One movement is that Earth travels around the

ARISTARCHUS 310-230 BCE

Sun, and the other movement is that Earth revolves around its own axis. We now know that he was right! But during his time no one believed him. It would be almost 2000 years before astronomers would look closely at his ideas.

1.3 Famous Early Astronomers

Nicolaus Copernicus was a famous astronomer who also thought that the Earth moved around the Sun. Copernicus was born in 1473 in the ancient Polish city of Torun. During the time Copernicus lived, most scientists believed that the Sun revolved around the Earth. They believed that the Earth was the center of the universe and everything revolved around it.

COPERNICUS 1473–1543 CE

Copernicus did not agree with the scientists of his day. His ideas would eventually change the whole science of astronomy! Unlike Aristarchus, Copernicus was able to use mathematics to show that the Earth moves around the Sun and that the Sun remains fixed in one location. However,

Copernicus was not outspoken about his ideas. Because he knew his ideas might upset people, he didn't talk about them. When Copernicus did publish his work, a few people got upset, but most people just ignored his hard work. Another 100 years passed before people took his ideas seriously.

BRAHE 1546-1601 CE

Another famous astronomer also changed the way we see the cosmos. His name was Tycho Brahe. Brahe was born in 1546 in the Danish town of Scania, and he was raised by his uncle. Like Copernicus, Brahe was curious about astronomy. His uncle wanted him to be a lawyer or a politician, but Brahe studied mathematics and slipped away at night to look at the sky. When his uncle died, Brahe was free to pursue his interests in astronomy.

Telescopes that make faraway objects look closer were not yet invented, so Brahe used sighting tubes, which are just hollow tubes with no lenses. Sighting tubes can be used to look at one star at a time. In this way, Brahe discovered that stars do not always appear to be in the same position but are constantly changing. Based on his observations, Brahe decided to rewrite the map of the stars and spent his life working on his ideas.

Galileo Galilei was also a famous early astronomer. He was interested in trying to find out how the planets move. Galileo was born in 1564 in Pisa, Italy. He studied many different subjects, such as mathematics and physics, and he loved to look at the stars. Galileo used his knowledge of math and physics to better understand how the planets and the Moon move.

GALILEO 1564-1642 CE

Like Copernicus, Galileo was an independent thinker, and he didn't believe in an Earth-centered universe. Galileo did experiments because he wanted to show how things moved rather than just coming up with ideas about it. By doing experiments and by using mathematics and physics, Galileo was able to prove that we live in a Sun-centered solar system that is made up of the Sun and the objects traveling around it. Being able to prove an idea by using experiments, math, and physics was the beginning of astronomy as a science.

1.4 Astronomers Today

Today, many scientists study the stars and planets. Astronomy is a science, and modern astronomers are scientists who use a variety of scientific tools and scientific techniques to learn about the universe.

However, even with new tools, modern astronomers must use the same basic skills that Copernicus, Brahe, and Galileo used.

Today's astronomers must make good observations and must train themselves to see the details, like Copernicus did. Astronomers must also study math and physics like Brahe and Galileo did. Math and physics are essential for understanding how the stars and planets move in space. Most importantly, astronomers must always be curious and willing to argue to defend their ideas like Copernicus, Brahe, and Galileo did.

1.5 Summary

- Astronomy is the study of space and all the objects found in space.

- Early astronomers were able to discover a great deal about the stars and planets by using observation.

- Nicolaus Copernicus, Tycho Brahe, and Galileo Galilei were three early astronomers who changed the way we understand the universe.

- Modern astronomers still use observation, math, and physics to study space.

1.6 Some Things to Think About

- When it gets dark, go outside and look at the sky. What can you notice? If you have binoculars, look through them and see if you can notice more details.

- If you notice the position of the Sun at different times during the day, does it look like the Sun moves around the Earth or the Earth moves around the Sun? Do you think events can sometimes be different than how they appear to be?

- Why do you think the new ideas of Copernicus and Brahe were upsetting to people?

- Think about times when you have seen the Moon in the sky. What things would you like to find out about the Moon?

Chapter 2 Astronomer's Toolbox

2.1   Introduction   11

2.2   Brief History of a
        Basic Tool   11

2.3   What Is a Telescope?   13

2.4   Early Telescopes   14

2.5   Advanced Telescopes   15

2.6   Summary   16

2.7 Some Things to Think Abbut   16

2.1 Introduction

How can you study objects in space when they are so far away? How can you make discoveries about what other solar systems are like when you can't get on a spaceship and go to them? How can you find out what other planets are like when you can't land on them?

Astronomers depend on a variety of tools to help them visualize the cosmos. New discoveries in astronomy have been made possible by the invention of computers, satellites, and spaceships that can travel far from Earth. Although the study of astronomy is changing quickly as more new tools help astronomers explore space, the invention of a basic tool, the telescope, was the first step in the exploration of space.

2.2 Brief History of a Basic Tool

In the 1400s and 1500s when Copernicus was gazing at the night sky, looking up at the stars and wondering about the cosmos, he had to do all of his investigating without the help of a telescope.

Copernicus was smart, and he revolutionized the way we think about the universe by using mathematics to show that the Earth moves around the Sun and that the Sun remains fixed in one location. But without being able to see the details of the Sun and of planets and their moons, Copernicus was limited in what he could explore.

Galileo was exploring the skies, just like Copernicus, but Galileo was born during the time when the telescope was being invented. In the early 1600's Galileo found out about a spyglass invented by a Dutch lens maker named Hans Lippershey. Galileo knew that this spyglass could help him study the cosmos by allowing him see things that were far away in more detail.

Galileo used the ideas that Lippershey had come up with and created his own spyglasses that would eventually become telescopes.

2.3 What Is a Telescope?

A telescope is a tool or instrument that helps astronomers see far into the distance. A simple telescope has two lenses connected by a long tube.

One lens is called the eyepiece and is located at one end of the tube. You look through this lens with your eye.

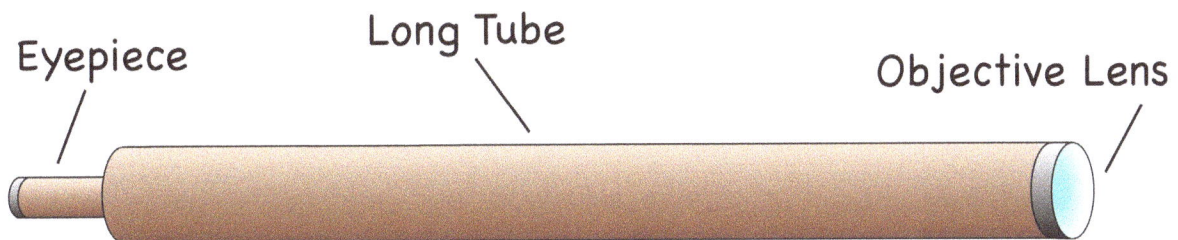

Eyepiece   Long Tube   Objective Lens

The other lens is called the objective lens and is located at the other end of the tube. The objective lens is used to collect light so an object may be viewed.

As light travels through the objective lens and down the tube, it is focused by the eyepiece. In this way the object being viewed becomes magnified, or made to appear larger. The larger the objective lens, the more light can be collected, and the longer the tube, the larger the object will appear.

2.4 Early Telescopes

Early telescopes looked very much like the simple telescope described in the previous section. Early telescopes were often made of metal, such as brass or copper, and had two simple lenses, one at either end of a long tube.

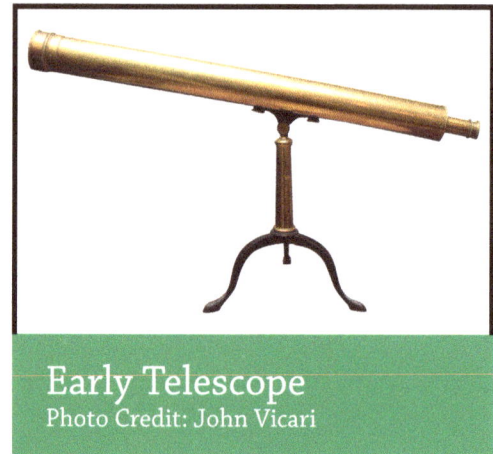

Early Telescope
Photo Credit: John Vicari

The body of the metal tube was sometimes attached to a stand. The astronomer could pivot the telescope on its stand, adjusting the direction in which the telescope was pointing.

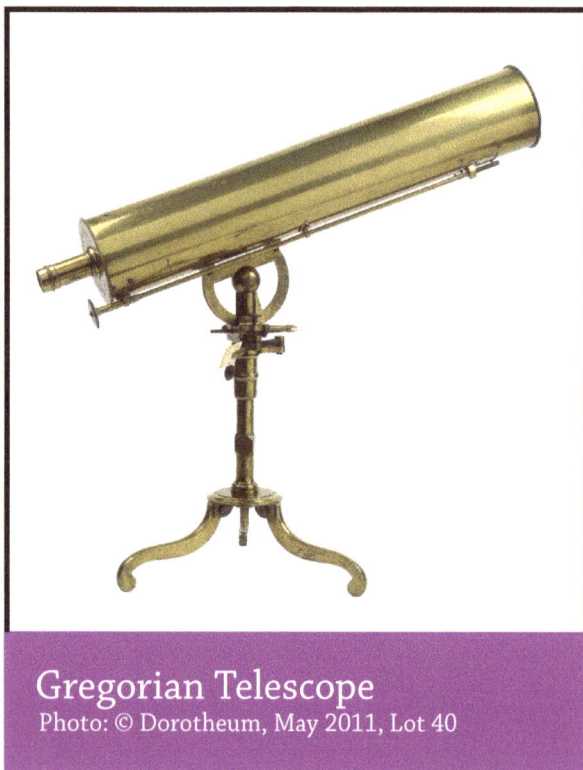

Gregorian Telescope
Photo: © Dorotheum, May 2011, Lot 40

Part of the problem with these early telescopes was that in order to see a faraway object the tube had to be very long. With the invention of the Gregorian telescope and the Newtonian telescope, mirrors were added to the inside of the telescope tube to solve this problem.

In both the Gregorian and Newtonian telescopes, mirrors in the tube are used to focus the light that enters through the objective lens. When the light hits the mirrors, it is bounced from the mirrors into the eyepiece. This design allows astronomers to see distant objects with telescopes that have shorter tubes.

2.5 Advanced Telescopes

The telescope has come a long way from the spyglass of Galileo's day. Technological advances have resulted in significant improvements for the modern telescope. Today, astronomers can see not only the planets and their moons in our own solar system but also planets and moons in other solar systems!

A galaxy is a large group of solar systems, stars, and other objects in space. Using modern telescopes, astronomers can now see whole solar systems, not only in Earth's galaxy, but in other galaxies as well!

The Hubble Space Telescope is an advanced telescope. As the name suggests, the Hubble Space telescope is actually in space orbiting the Earth. The Hubble Space Telescope was launched into space in 1990 and has produced some of the most amazing images of space we have ever seen.

2.6 Summary

● Astronomers explore space by using tools such as telescopes, computers, satellites, and spaceships.

● Galileo modified the spyglass to create a telescope.

● Early telescopes had an objective lens and an eyepiece that were connected by a long tube.

● The Hubble Space Telescope is an advanced telescope that is in space and travels around the Earth.

2.7 Some Things to Think About

● What have you observed through a telescope?

The Moon
Jupiter
Venus
Galaxies
A comet
Something else

● What do you think it was like for Galileo when he first looked at the cosmos through a telescope?

● How do you think you would build a simple telescope?

● How do you think the invention of Gregorian and Newtonian telescopes helped astronomers?

What images have you seen from the Hubble Space Telescope? Which were the most surprising?

galaxies

nebulae

comets

planets

other stuff in space

Chapter 3 Earth's Home in Space

3.1 Introduction   19

3.2 Earth Is a Planet   20

3.3 The Moon and Tides   23

3.4 The Sun and Weather   24

3.5 Eclipses   25

3.6 Summary   25

3.7 Some Things
     to Think About   26

COWABUNGA!

3.1 Introduction

Now that we know what astronomy is and how to study planets and stars, it's time to explore what Earth looks like from space.

Because Earth is so big compared to our human size, it's hard to imagine what Earth looks like from space. Is Earth the biggest object in space? Is Earth in the center of space? If we took a rocket into space, what would we find?

3.2 Earth Is a Planet

If you launched a rocket and traveled past the clouds into space, you would see the Earth. Earth would look like a blue marble floating in the dark space around it.

Earth is a planet. A planet has special properties. A planet has to be large enough to have its own gravity, which is the force that holds everything to the Earth's surface. A planet also has to move in space around a sun. And finally, a planet is shaped like a ball. Spherical is the word for ball-shaped.

Because Earth is very large, moves around the Sun in space, has gravity, and is spherical, Earth is called a planet.

Earth rotates around an axis, which is an imaginary line that goes through the center of an object. If you were to take a ball and spin it with your fingers, it would rotate around an axis.

The Sun shines on different parts of Earth as Earth rotates. This is how we get our days and nights.

It takes 24 hours for Earth to rotate once around its axis. We don't feel the rotation of the Earth because the Earth's gravity holds down the air and everything else that's on Earth. Everything is moving at the same speed, and the Earth's rotation does not cause wind.

Earth is tilted on its axis. "Tilted" means that Earth is not just straight up and down but is slanted. The tilt of Earth's axis gives us seasons.

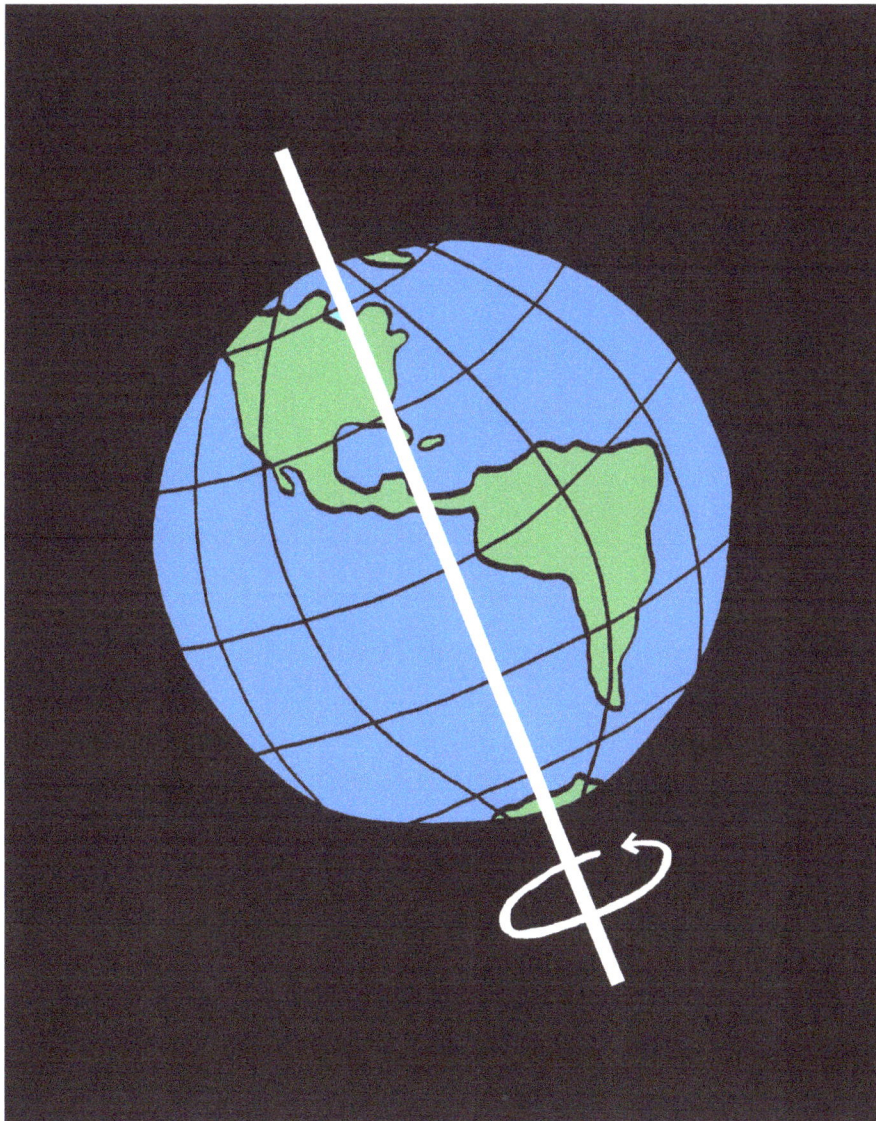

For part of the year, the northern part of Earth is tilted towards the Sun and the southern part of the Earth is tilted away from the Sun. This gives the northern part of the Earth summer and the southern part winter.

Then, during a different part of the year, the southern part of the Earth is tilted towards the Sun and the northern part away from the Sun. When this happens, the southern part of Earth has summer, and the northern part has winter.

3.3 The Moon and Tides

If you take your rocket into space, you might run into the Moon. A moon is an object that travels around a planet. Our moon is smaller than Earth and travels around Earth.

Our moon does have some gravity and is spherical, but because it moves around Earth and not around the Sun as the Earth does, the Moon is not a planet. We will learn more about the Moon in the next chapter.

Did you know that the Moon helps create ocean tides on Earth? It's true! The Moon pulls on Earth with some gravity. This pulling on the Earth causes the water in the oceans to be pulled too.

As the Moon moves around the Earth, it pulls the ocean water with it. The pulling of ocean water by the Moon helps create tides.

3.4 The Sun and Weather

If you wear sandals during a summer day, you can feel the Sun warming your toes. The Sun is a big ball of fire that gives light and heat energy to Earth.

The Earth orbits the Sun. The word orbit means to "revolve around." An orbit is the path one object makes as it travels around another object. If you stand with your hand on a pole and then start walking, you will make a path around the pole. You will orbit the pole. This is what Earth does as it moves around the Sun. We will learn more about the Sun in the next chapter.

Did you know there are storms on the Sun? Did you know that the storms on the Sun can cause storms on Earth? It's true! Sun storms can contribute to Earth storms. Scientists who research weather can study Sun storms to find out how they affect Earth's weather.

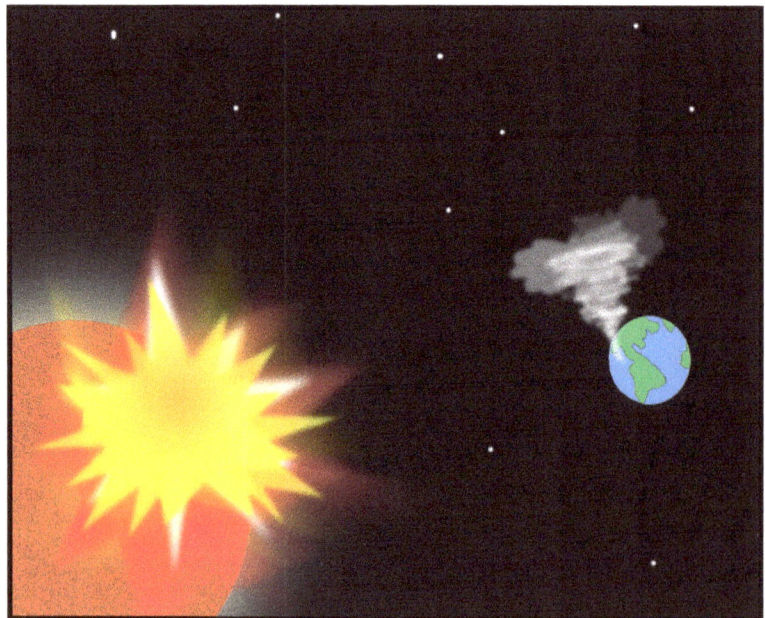

3.5 Eclipses

During the Moon's orbit around Earth, the Moon travels behind the Earth. When the Earth is in between the Moon and the Sun, the Earth can block the Sun's light from reaching the Moon. When the Earth's shadow is cast on the Moon, it is called a lunar eclipse.

At other times, the Moon will be in between the Sun and the Earth. With the Moon in this position, the Moon can block the Sun's light from reaching a portion of the Earth. This is called a solar eclipse.

3.6 Summary

- Earth is a planet.

- One rotation of the Earth on its axis takes 24 hours (one day).

- Earth is tilted on its axis, giving us seasons.

- The Moon and Sun affect Earth's tides and weather.

3.7 Some Things to Think About

● If you were on the next rocket that was launched into space, what do you think you would see? What would you like to see?

● What do you think would happen if Earth's axis changed from being tilted to being straight up and down?

● What do you think it would be like if Earth did not have a Moon?

● Which do you think would be most interesting to study? Why?

Sun storms
Earth's weather
Earth's orbit around the Sun
The Moon and tides
Earth's tilt on its axis

● If you were watching a lunar eclipse, what do you think you might see?

Chapter 4 Earth's Neighbors: Moon and Sun

4.1   Introduction   28

4.2   The Moon   28

4.3   The Sun   30

4.4   The Sun's Energy   32

4.5   Summary   33

4.6   Some Things
      to Think About   34

4.1 Introduction

In the last chapter you saw that the Earth sits in space. You also saw that the Moon and Sun cause changes in our ocean tides and weather. But what is a moon and what is a sun?

4.2 The Moon

You can see the Moon from Earth. If you look outside your house, you might see a bright, round shape in the sky. This is the Moon.

You can sometimes see the Moon during the day, but most often you see the Moon as the brightest object in the sky at night.

The Moon is spherical like the Earth but much smaller. The Moon has a very different surface from that of Earth. Although the Moon is made of rocks and minerals like Earth, the Moon cannot support life. It has very little oxygen and no liquid water.

The Moon looks bright in the sky, but the Moon does not generate its own light. Acting like a mirror, the Moon reflects the Sun's light to Earth.

The Moon orbits Earth once a month. If you look at the Moon often, you will see that it appears to change its shape during the month. The different shapes are caused by different views of the Moon as we see it from Earth. The Moon can be round, half-round, or crescent shaped. When the Moon looks round, we say it is a full moon.

If you look closely at the Moon, you can see light and dark patches. These light and dark patches sometimes look like faces. The "Man in the Moon" is a famous nursery rhyme from Mother Goose.

The light and dark patches on the Moon are actually craters and lava flows.

4.3 The Sun

The other large object you see in the sky is the Sun. You can't look directly at the Sun (that would damage your eyes), but you can see that the Sun is rising in the morning, moving across the sky during the day, and setting at night.

The Sun is not a planet or a moon. The Sun is a star. A star is any object in space that generates its own light and heat energy (see the next section).

The Sun is much larger than Earth. The Sun is so large that a million Earths would fit inside!

The Sun is also very hot. The temperature on the Sun's surface is thousands of times hotter than temperatures on Earth. The Sun is so hot that it would melt everything on Earth if Earth were too close to it!

THE SUN   Photo credit: nasaimages.com

4.4 The Sun's Energy

The Sun generates its own light and heat energy. From physics we know that energy is something that gives something else the ability to do work. The energy made by the Sun comes out as light energy and heat energy. Plants use the Sun's light energy to make food. Making food is work, so the Sun gives plants the ability to do work. We can use the Sun's light energy to make electricity. Our bodies also use the Sun's heat energy to stay warm. Without the Sun there would be no life on Earth.

The Sun is not made of rocks and is not like the Earth and the Moon. Instead, the Sun is made of helium and hydrogen gases. The extremely hot temperatures on the Sun make hydrogen atoms stick together, forming helium atoms.

From chemistry you know that when atoms stick together they can make molecules. You have also learned that in a chemical reaction atoms and molecules rearrange to make new molecules. When atoms stick together to make molecules, they use their electrons to attach to each other.

When hydrogen atoms stick together to make helium atoms, they rearrange their protons and neutrons! This is called a nuclear reaction. A nuclear reaction is different from a chemical reaction. In a chemical reaction, the protons and neutrons stay the same and only the electrons stick together.

During a nuclear reaction lots of light and heat energy are released. Because the Sun can generate its own energy, it is called a star.

4.5 Summary

● The Moon orbits the Earth once a month.

● The Moon is made of rocks and minerals, like Earth.

● Our Sun is a star and makes its own energy.

● Our Sun is made of helium and hydrogen gases.

4.6 Some Things to Think About

● Do you think the Moon would look different if it had liquid water and plants growing on it? Why or why not?

● What do you think life on Earth would be like if the Sun were only half as big as it is now? What do you think life on Earth would be like if Earth were twice as far away from the Sun as it is now?

● What are some ways in which the Sun and the Moon are different from each other?

Chapter 5 Earth's Neighbors: Planets

5.1   Introduction   36

5.2   Planets   36

5.3   Two Types of Planets   38

5.4   Where's Pluto?   41

5.5   Summary   42

5.6   Some Things
       to Think About   42

5.1 Introduction

In Chapter 4 we learned about two of Earth's neighbors—the Moon and the Sun. In this chapter we will learn about some of Earth's other neighbors—the planets.

5.2 Planets

If you look into the sky during the day, you can see the bright Sun and possibly a very faded Moon.

But at night, depending on where you live, the sky twinkles up with brilliant specks of light. Many of these specks are stars, like our Sun. But some of the bright lights that dot the night sky are planets.

In Chapter 3 we found out that Earth is a planet. Recall that Earth is a planet because it is large enough to have gravity, is spherical in shape, and orbits the Sun. The Moon is not a planet because it orbits the Earth rather than the Sun.

Earth is one of eight planets that orbit the Sun. The names of these eight planets are: Mercury, Venus, Earth, Mars, Jupiter, Saturn, Uranus, and Neptune.

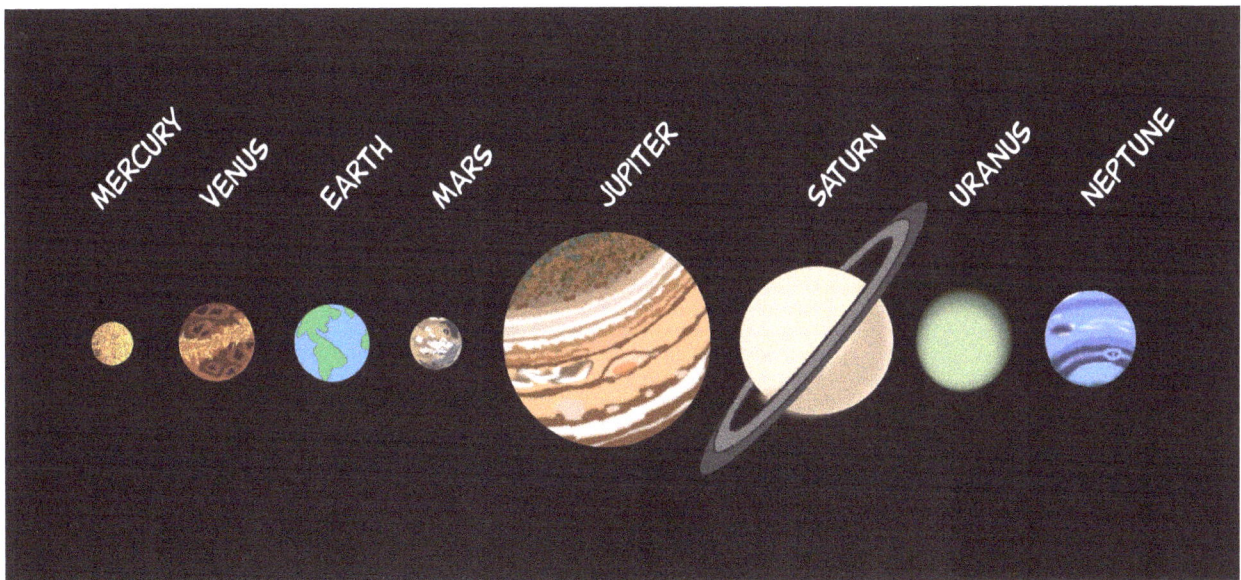

5.3 Two Types of Planets

All of the planets are different from each other. Mercury has a barren moon-like surface, Venus has toxic gas clouds, and Saturn has brilliant rings. Mars has a reddish color, and Uranus and Neptune look blue or blue-green.

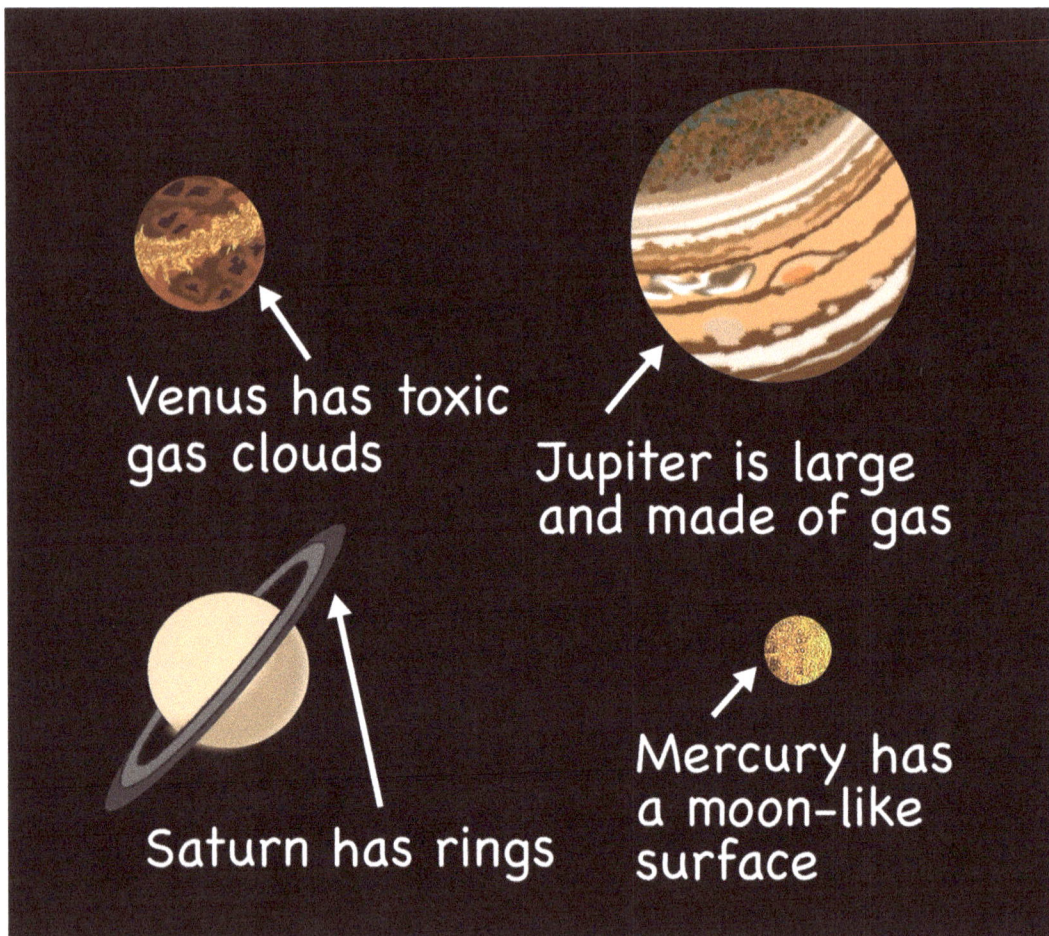

Venus has toxic gas clouds

Jupiter is large and made of gas

Saturn has rings

Mercury has a moon-like surface

Even though all the planets are different from each other, some of the planets have features that are similar. Because of these similarities, scientist are able to separate the planets into two groups. The names of these two groups are terrestrial planets and Jovian planets.

The terrestrial planets are those planets that are most like Earth, and the word terrestrial means "Earth-like." There are four terrestrial planets: Mercury, Venus, Earth, and Mars.

All of the terrestrial planets are made of rock and minerals, like Earth. Also, all of the terrestrial planets have volcanoes, mountains, and craters on their surfaces.

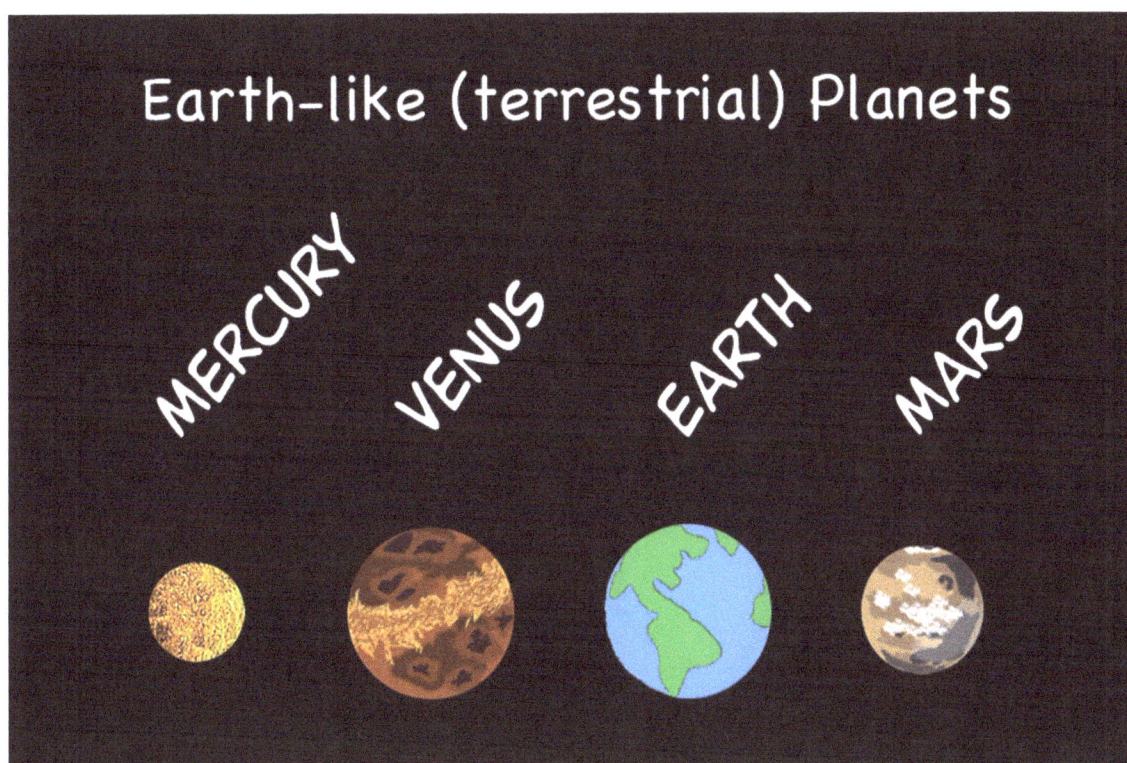

Earth-like (terrestrial) Planets

MERCURY VENUS EARTH MARS

However, of all the terrestrial planets, only Earth has life on it. Only Earth has the water, the oxygen, and the proper atmosphere needed to support life as we know it.

The Jovian planets are those planets that are similar to Jupiter. Jupiter is a very large planet made mostly of hydrogen gas and helium gas. All of the Jovian planets are like Jupiter because they are all very large and made mostly of gas. The Jovian planets are Jupiter, Saturn, Uranus, and Neptune.

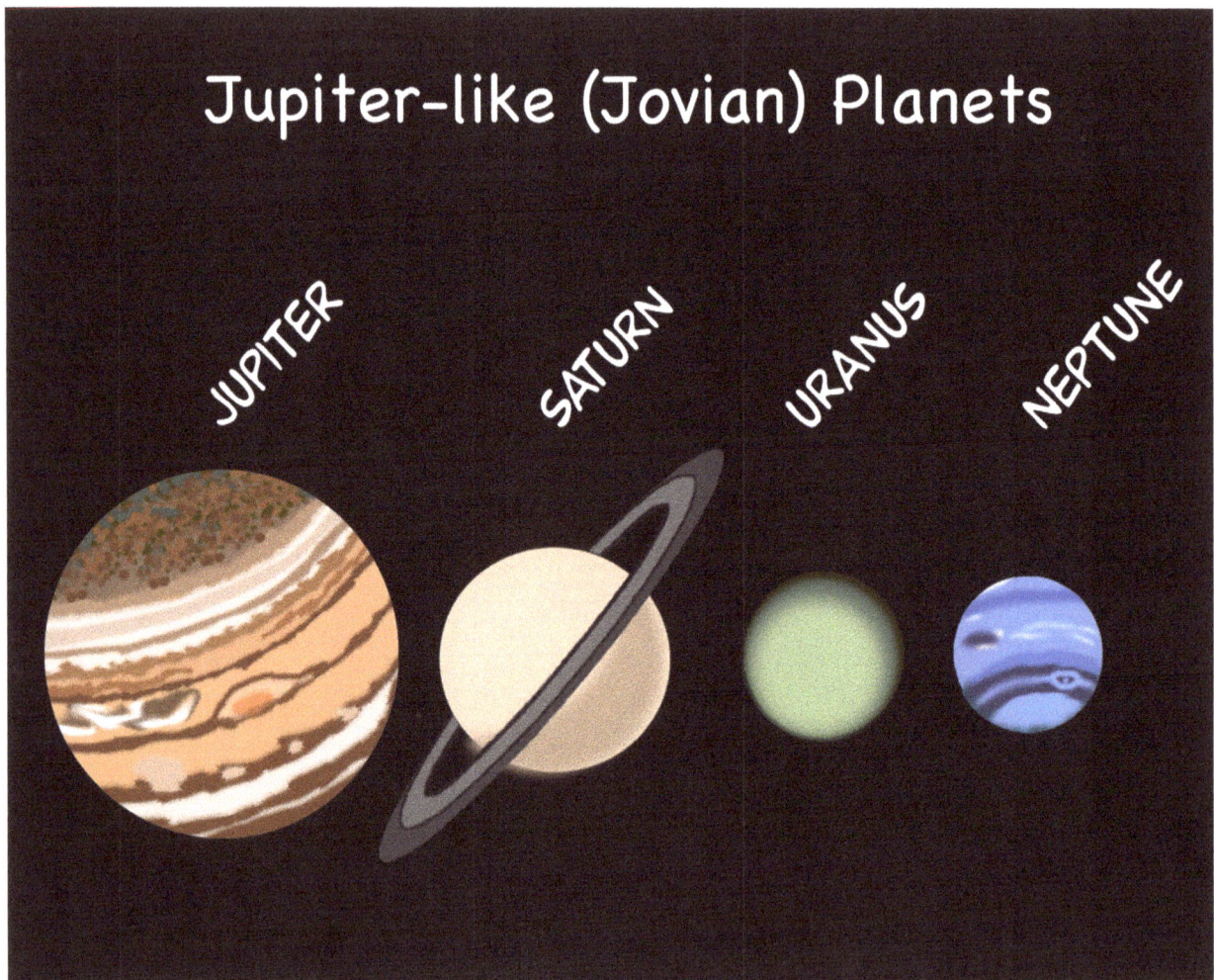

Jupiter-like (Jovian) Planets

JUPITER   SATURN   URANUS   NEPTUNE

5.4 Where's Pluto?

Pluto was once called the 9th planet, but in August 2006 scientists at the International Astronomical Union (IAU) changed their minds. They decided Pluto does not have all the features needed to be classified as a planet. Pluto is now called a dwarf planet or a plutoid.

However, not all scientists agree with the IAU's decision. Scientists argue about conclusions all the time, and arguing is part of science. Someday Pluto may again be considered a planet.

5.5 Summary

- The eight planets are Mercury, Venus, Earth, Mars, Jupiter, Saturn, Uranus, and Neptune.

- The planets are separated into two groups: terrestrial planets and Jovian planets.

- Terrestrial planets are "like Earth." Mercury, Venus, Earth, and Mars are terrestrial planets.

- Jovian planets are "like Jupiter." Jupiter, Saturn, Uranus, and Neptune are Jovian planets.

5.6 Some Things to Think About

- We know that Earth has a moon orbiting it. Do you think any of the other planets in our solar system might have a moon? Do you think a planet could have more than one moon?

● If you could travel in space, which planet would you most like to visit? Why?

Mercury

Venus

Mars

Jupiter

Saturn

Uranus

Neptune

I would stay on Earth.

● Why do you think arguing about ideas and conclusions can be an important part of science?

Chapter 6 Observing Constellations

6.1 Introduction   45

6.2 Northern Hemisphere
     Constellations   47

6.3 Southern Hemisphere
     Constellations   49

6.4 Using Stars to Navigate   50

6.5 Summary   53

6.6 Some Things
     to Think About   53

6.1 Introduction

No matter where you are on Earth, if you look up to the sky on a clear, dark night, you can see stars. If you are far away from city lights, it looks like the sky is filled with thousands and thousands of stars. There are so many stars that it is difficult to know the name of each star in the sky.

Even without using a telescope, modern satellites, or space probes, we can learn something about the cosmos by observing stars that form constellations. A constellation is a group of stars that together appear to form a shape or image in the sky. By grouping stars into constellations, a lot can be learned about the sky without knowing all the individual stars.

Some common constellations easily observed in the Northern Hemisphere include the Big Dipper and Cassiopeia. Favorite Southern Hemisphere constellations include the Archer and the Whale.

Since Earth is a large sphere, the stars someone can see at the North Pole are different from the stars someone can see at the South Pole. If you are at the equator, over the course of a year you can see all the constellations, but the stars at the North and South Poles will be harder to see.

The constellations change their positions in the sky throughout the night and during the different seasons. This repositioning of the stars is caused by the Earth changing its position in space as it spins on its axis and revolves around the Sun. As the Earth moves, we see the constellations from different angles.

MY INSTRUMENTS ARE OUT!

NO WORRIES. I SEE THE BIG DIPPER SO I KNOW EXACTLY WHERE WE ARE.

By observing the positions of the constellations in the sky, you can watch the seasons come and go, tell what time of night it is, and determine which direction you are going when traveling on land, sea, or in the air. Knowing the constellations could even help you find your way home if you were taking a trip in outer space!

6.2 Northern Hemisphere Constellations

If you live in the Northern Hemisphere there are many constellations you can easily find. A favorite group of stars in the northern sky is called the Big Dipper because its shape looks like a dipper, or ladle. The Big Dipper has 7 stars, three of which form the "handle" and four that form the "bowl" of the dipper. The best time to see The Big Dipper is from February through June.

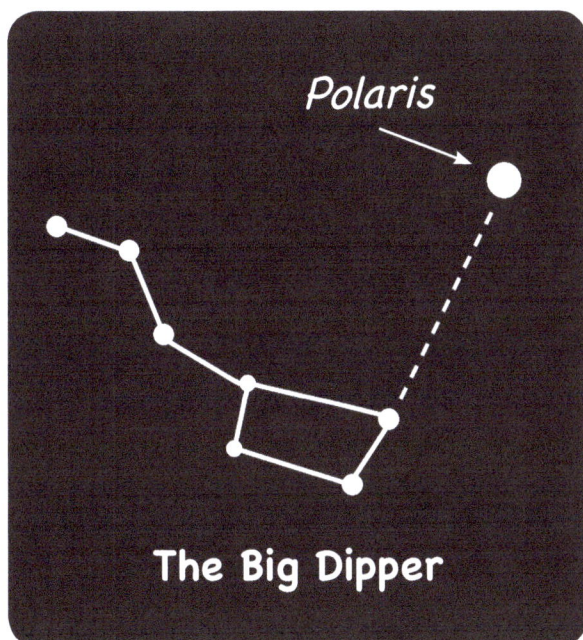

Polaris

The Big Dipper

The Big Dipper can be used to locate the star Polaris, which is also called the North Star. Polaris can be found by imagining a line going between the two stars located at the end of the bowl of the Big Dipper and extending the line to a lone star that has no other stars nearby. This star is Polaris.

Polaris is above the northernmost point of the Earth. Because it is near enough to true north, it can be used as a navigational marker. If you were on a boat sailing from Maine to England on a clear night, you could use Polaris to keep your boat on track.

Big Dipper

The Great Bear

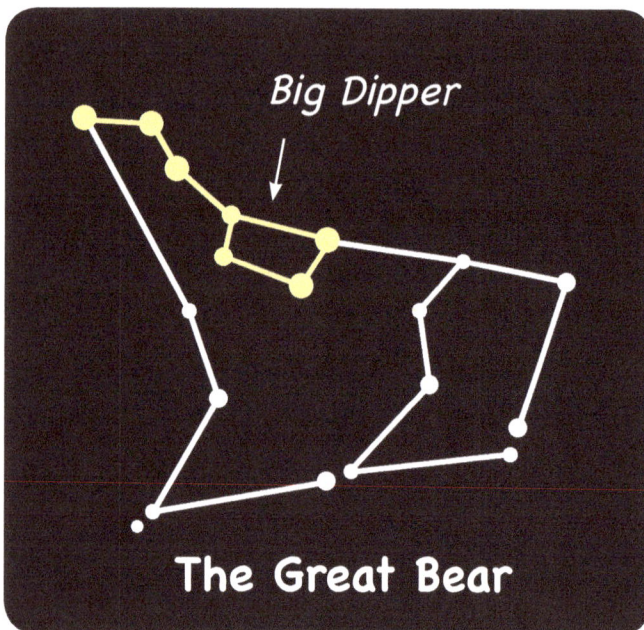

Another interesting constellation to find in the Northern Hemisphere is the Great Bear. The Great Bear includes the Big Dipper. To locate the Great Bear, first find the Big Dipper, and while keeping the Big Dipper in view, expand your gaze to include the three pairs of stars that form the Bear's paws. Once you have these stars and the Big Dipper in view, you can see the rest of the stars that make up the Great Bear. The best time to see the Great Bear is from February through June.

Another favorite constellation is Orion the Hunter. The stars that make up Orion are bright and beautiful which makes it easy to pick out this constellation. A good way to find Orion is to find his belt. Orion's belt is made of three bright stars close together in a straight row. From there you can pick out the shield and raised club. The best time to see Orion is December through March.

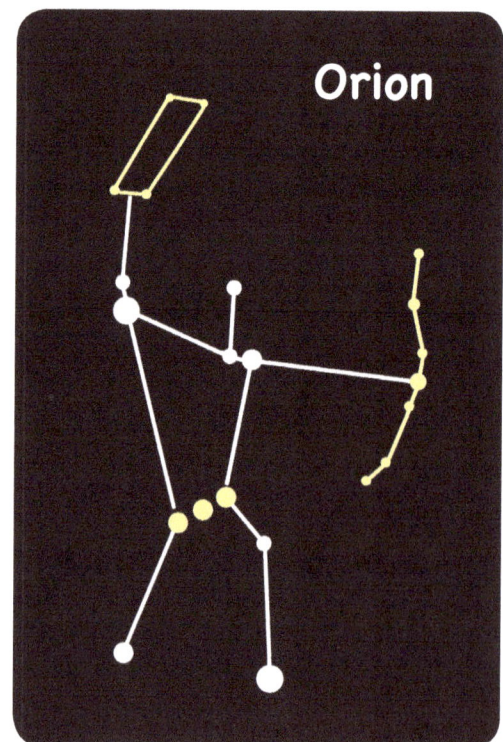

Orion

6.3 Southern Hemisphere Constellations

If you live in the Southern Hemisphere, there are lots of fun constellations you can find. One of the largest southern constellations is the Whale. The stars that make up the Whale are dim, but because there are fewer stars to observe in this section of the sky, the Whale can be easily seen on a dark night when there is no Moon or city lights.

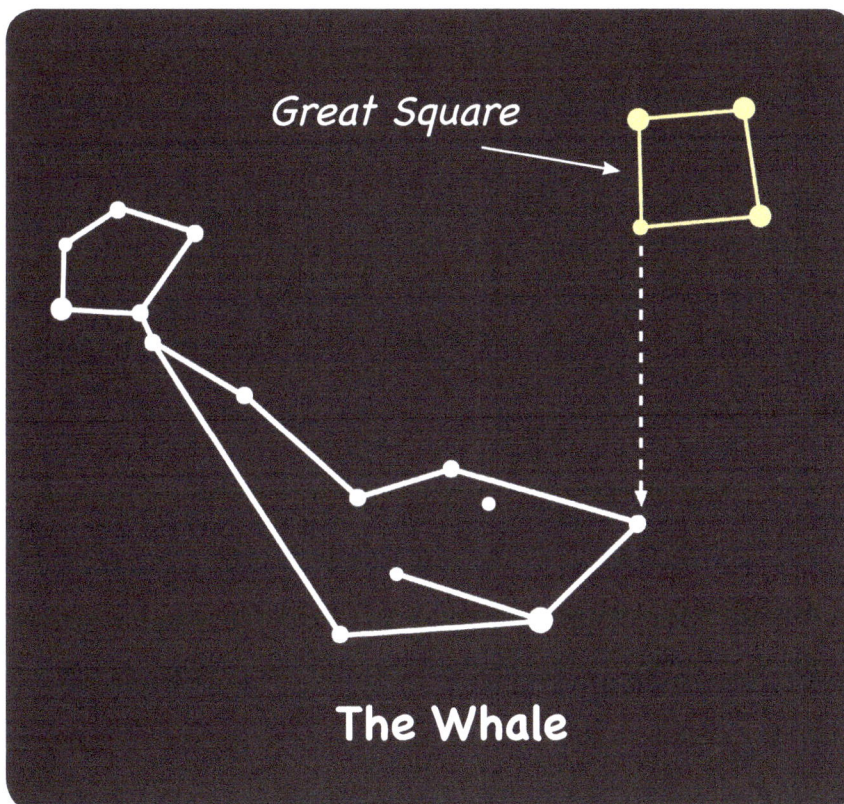

Great Square

The Whale

To find the Whale, first locate the Great Square. As the name suggests, the Great Square is a set of four bright stars that form a square. Once you locate The Great Square, the Whale is easy to spot. Just follow the line made by the two stars on the side of the Great Square until you see a cluster of 4 to 5 stars. These are the stars that make up the head of the Whale. The best time to see the Whale is October through January.

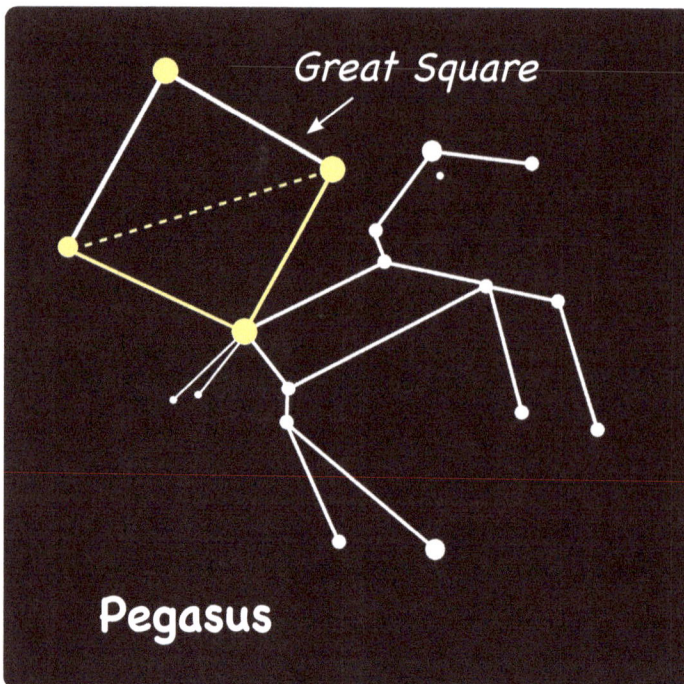

Great Square

Pegasus

Another favorite constellation in the Southern Hemisphere is Pegasus, the winged horse. Part of the Pegasus constellation includes three of the four stars of the Great Square. These three stars make up the wing of Pegasus which sits on the hind end of the horse. Two little stars near the hind end make up the tail and the head extends in the opposite direction from the tail. The best time of the year to see Pegasus is from August to October.

6.4 Using Stars to Navigate

How do you find your way to the grocery store? How do you know which street to take to the park? If you need to go to a friend's house, do you turn to your right, left, or go straight ahead from your front door?

In each of these situations, you are navigating your way from one place to another place. Navigation simply means to make one's way from one location to another. There are several different techniques people use to navigate.

One way to navigate is to use landmarks. If you walk to the grocery store with your parents, you might notice that it is located just across from a park and next to a gas station. The next time you need to go to the grocery store, you can use your knowledge of the park and the gas station as landmarks to guide you to the store. However, if you were far out to sea where there are no landmarks, how would you find your way home?

This was a problem for early sailors. When traveling along the coast, they could use landmarks to find their way, but what happened when they traveled far enough out in the ocean that they could no longer see the shore? Early sailors discovered that they could use the stars as a way to navigate across the sea. Using the stars is

a great way to find your way, whether you are on land, sea, in an airplane or even in space!

One easy star to use for navigation in the Northern Hemisphere is Polaris. Unlike other stars, Polaris actually stays in the same place in the sky and doesn't appear to move. The north pole of Earth's axis points almost directly at Polaris. If you can find Polaris, you can tell which way is north, and once you locate north, you can find south, east, and west.

In the Southern Hemisphere people use the Southern Cross constellation for navigation. Although Earth's axis at the South Pole doesn't point directly at an individual star, the two stars that form the long part of the constellation can be used to find south.

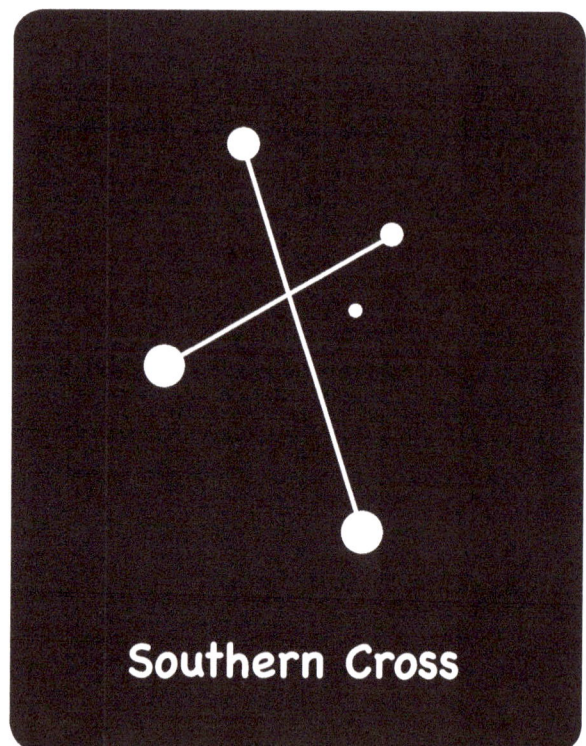

Southern Cross

6.5 Summary

- A constellation is a group of stars that together form a particular shape.

- Common constellations in the Northern Hemisphere include the Big Dipper, the Great Bear, and Orion.

- Common constellations in the Southern Hemisphere include the Whale and Pegasus, both of which can be located by finding the Great Square.

- Stars can be used to navigate, or find the way, from one location to another.

6.6 Some Things to Think About

- Why do you think would you see different stars from the North Pole than you would from the South Pole?

- Which is your favorite Northern Hemisphere constellation?

  The Big Dipper
  Orion
  The Great Bear
  Cassiopeia

- Which is your favorite Southern Hemisphere constellation?

  Pegasus
  The Southern Cross
  The Whale

- Have you ever used the stars to navigate? How would you do it?

Chapter 7 Earth's Neighborhood

7.1 Introduction   55

7.2 Our Solar Neighborhood   55

7.3 Orbits   58

7.4 Why Is Earth Special?   61

7.5 Summary   62

7.6 Some Things to Think About   62

7.1 Introduction

In our solar system there are two different types of planets. Recall that some of our planetary neighbors are Earth-like, or terrestrial, and some are Jupiter-like, or Jovian. In this chapter we will take a look at where our planetary neighbors "live" in our solar neighborhood.

7.2 Our Solar Neighborhood

Most people live in some kind of neighborhood. A neighborhood is an area of town with houses, apartments, a few businesses, and possibly a park.

If you take a walk down the block in your neighborhood, you can see where your neighbors live. Some of your neighbors live close to you. Maybe they live next door and share the same backyard. Other neighbors live farther away, but they may all go to the local grocery store or walk their dog in the local park. We would say that all of the people who live in this particular area of town are part of a neighborhood.

In the same way, planets share a particular area in space. A solar system is made up of a sun and the planets and other objects that travel around that sun.

In our solar system, there are eight planets. All of the planets share the same sun. Some planets are closer to our Sun, and some are farther away, just like the neighbors in your neighborhood.

The closest planet to the Sun and the smallest planet is Mercury. Because Mercury is so close to the Sun, its surface can be very hot. The temperature at noon on Mercury can get up to as much as 425 degrees Celsius (800 degrees Fahrenheit)! But Mercury does not have enough air to hold onto the heat from the Sun. At night the temperature on Mercury can go down to below 18 degrees Celsius (below zero degrees Fahrenheit). So Mercury does not have the right temperatures for plants and animals to be able to live.

The next closet planet to the Sun is Venus. Venus is about twice as far away from the Sun as Mercury. However, even though Venus is farther away from the Sun than Mercury is, Venus is actually hotter! Venus has lots of carbon dioxide in the air. This heats up the surface and holds the heat so

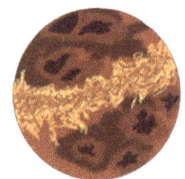

Venus is hot all the time. The surface of Venus can reach over 460 degrees Celsius (860 degrees Fahrenheit). Venus is much too hot to support plant and animal life.

The next closest planet to the Sun is Earth. Earth is close enough to the Sun to have enough heat for life to exist, but not so close that it is too hot for living things. Earth is the only planet in our solar system that supports plant and animal life.

Mars sits farther away from the Sun than Earth does. Mars is much colder than Earth because it is farther away from the Sun. However, Mars is almost close enough to the Sun to support life.

Mercury, Venus, Earth, and Mars make up the inner solar system, or inner neighborhood. From an astronomer's perspective, all of these planets are relatively close to each other.

Much farther out are the four planets in the outer solar system. Jupiter is the first planet in the outer solar system. Jupiter is more than five times farther away from the Sun than Earth is.

Saturn is even farther away from the Sun than Jupiter is. Saturn is the second planet in the outer solar system.

Uranus and Neptune are the last two planets in the outer solar system.

Neptune is almost 30 times as far away from the Sun as Earth is. Jupiter, Saturn, Uranus, and Neptune are all much too cold to support plant and animal life.

7.3 Orbits

The planets don't just sit in one spot, but move in a nearly circular orbit around the Sun. An orbit is a particular path, like a road, that a planet follows.

Each planet stays in its orbit at its particular distance away from the Sun. Planets don't cross other planetary orbits or ever bump into each other.

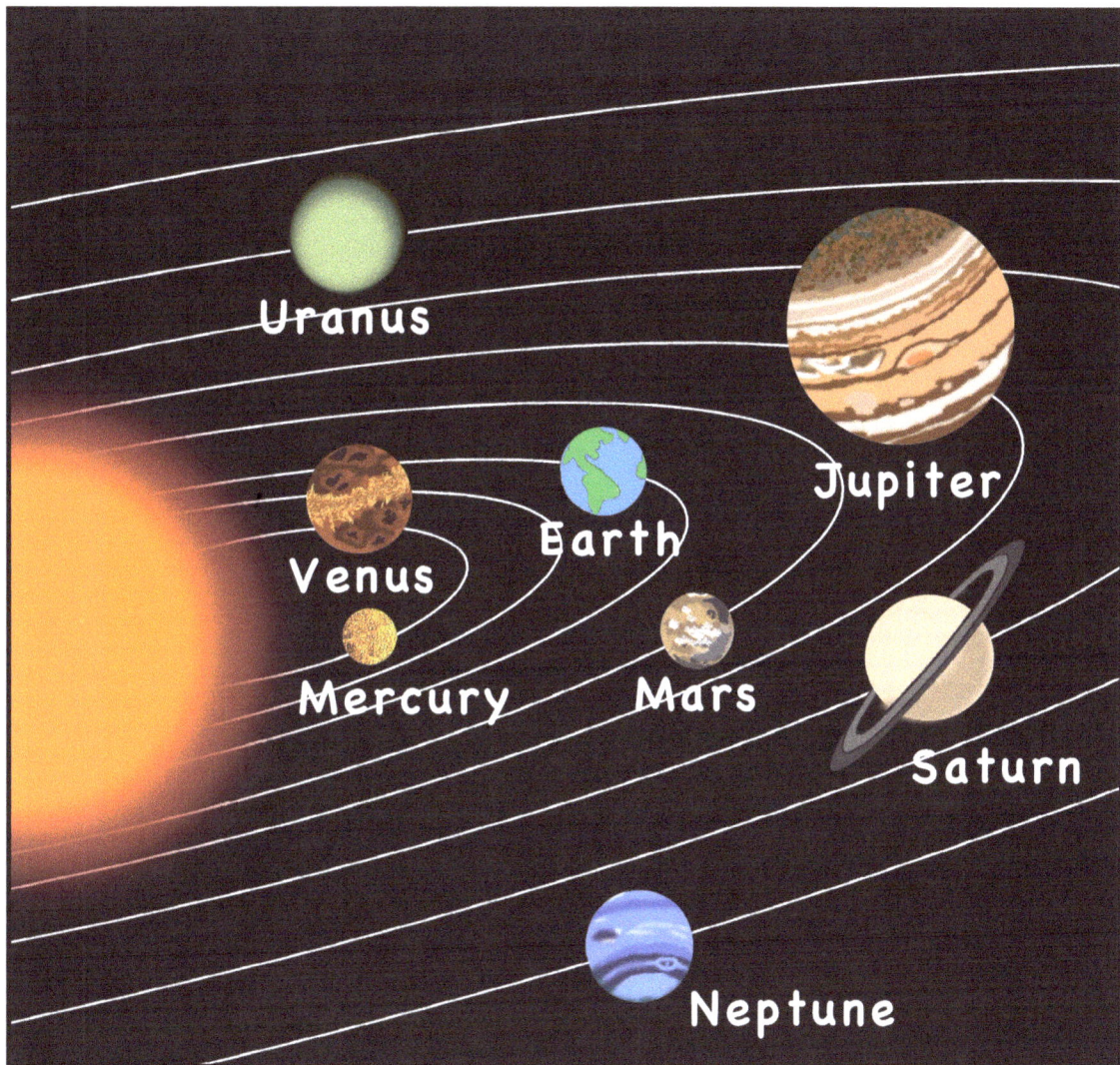

Each planet takes a certain number of days to orbit the Sun. This is called a planetary year. Mercury orbits the Sun faster than does any other planet. It only takes 88 days for Mercury to complete one orbit. So one year on Mercury is only 88 days.

It takes Venus a little longer than Mercury to orbit the Sun, but not as long as it takes Earth. Venus orbits the Sun in 225 days, and Earth orbits the Sun in 365 days. The length of our calendar year is 365 days.

It takes Mars 687 days to orbit the Sun, and it takes Jupiter almost 12 Earth years to complete one planetary year!

Saturn takes almost 30 Earth years to orbit the Sun, and Uranus orbits the Sun in 84 Earth years. If you lived on Neptune you would need 165 Earth years just to get around the Sun once!

Planet	Number of Days for One Orbit of the Sun
Mercury	88
Venus	225
Earth	365
Mars	687
Jupiter	4,332
Saturn	10,760
Uranus	30,700
Neptune	60,200

7.4 Why Is Earth Special?

Of all the planets in our solar system, only Earth is the right distance away from the Sun, with the right combination of water, oxygen, minerals, and soils to support plant and animal life. Earth has just the right conditions for plant and animal life to exist.

Earth has many unique features that make it just right for plant and animal life. If Earth were closer to the Sun, it would be too hot for life. If Earth were farther away from the Sun, it would be too cold for life. If Earth's atmosphere had too much gas, like Venus does, it would cause the Earth's surface to heat up and would make it too hot.

Earth is special in this way. There is no other planet in our solar system that can support plant and animal life. And so far, no other planet in the universe has been found that supports plant and animal life. Earth has just the right temperature, is just the right distance from the Sun, and is made of just the right materials for plant and animal life to exist!

7.5 Summary

● We live together with other planets in a planetary neighborhood called the solar system.

● All of the planets in our solar system share the same sun.

● Mercury is closest to the Sun, followed by Venus, Earth, Mars, Jupiter, Saturn, Uranus, and Neptune.

● Each planet rotates around the Sun in an orbit.

● Each planet takes a different number of days to complete one orbit around the Sun. The number of days it takes a planet to orbit the Sun once is called a planetary year. Our planetary year is 365 days.

7.6 Some Things to Think About

● If you could pick any planet in the solar system to live on, which one would it be? Why?

Mercury

Venus

Earth

Mars

Jupiter

Saturn

Neptune

Uranus

● What do you think it would be like to live on a planet like Jupiter? What do you think would be different? (holidays, birthdays, seasons?)

● If you could pick any planet in the universe to live on, what requirements would be needed for plants and animals to be able to live?

Chapter 8 Beyond the Neighborhood

8.1	Introduction	65
8.2	Nearest Star	65
8.3	Brightest Star	67
8.4	Biggest Star	68
8.5	Stars With Planets	69
8.6	Summary	70
8.7	Some Things to Think About	70

8.1 Introduction

In Chapter 7 we looked at our planetary neighborhood called the solar system. We saw how the eight planets orbit the Sun and why Earth is the only planet in our solar system that can support plant and animal life.

But what about other solar systems? What lies beyond our neighborhood? Are there other neighborhoods with suns like ours and planets like ours that support plant and animal life?

8.2 Nearest Star

If you look up into the sky on a dark, clear night, you can see lots of stars. On a moonless night, far away from the city lights, you might be able to see 2,000 stars or more.

The closest star to our Earth is a star called Proxima Centauri. Proxima Centauri is actually part of a three-star system. The two other stars in this system are called Alpha Centauri A and Alpha Centauri B. Alpha Centauri A, Alpha Centauri B, and Proxima Centauri orbit each other.

OUR SUN   PROXIMA CENTAURI   ALPHA CENTAURI A   ALPHA CENTAURI B

Proxima Centauri is smaller and cooler than our Sun and is red in color. Even though Proxima Centauri is the closest star to our Sun, it is not bright enough to be seen without a telescope.

8.3 Brightest Star

The brightest star we can see in the sky is called Sirius. Sirius is farther away than Proxima Centauri but is brighter. The brightness of a star does not depend on how close it is, but on how much light the star makes.

One way to think about how much light a star makes is to look at the difference between a candle and a flashlight. A candle will light up or illuminate an area of a meter or so (a few feet) around it, but a flashlight can illuminate several meters (yards). A flashlight puts out more energy than a candle and as a result can illuminate a much longer distance.

If you were to look at a candle and a flashlight from several meters (yards) away, the flashlight would look brighter. If you moved the flashlight a few meters (yards) farther away than the candle, it would still look brighter.

In the same way, Sirius is a brighter star than Proxima Centauri even though it is farther away.

8.4 Biggest Star

The biggest star is not necessarily the closest star or the brightest star. The biggest star we can see from Earth is called VY Canis Majoris or VY CMa.

VY CMa is about 2,000 times larger than our Sun. If VY CMa were in our solar system replacing our Sun, it would extend beyond the orbit of Saturn!

8.5 Stars With Planets

Scientists think that most stars have planets orbiting them.

It is very difficult for astronomers to see the planets in other solar systems directly. Planets are very dim because they don't produce their own light like suns do. However, astronomers can use other techniques to figure out if stars have planets orbiting them. One technique is to use physics to determine if a star is wobbling. As planets orbit a sun, their gravity pulls on the sun, causing it to wobble. Astronomers can study a sun's wobble to determine if one or more planets might exist in orbit around the sun.

Many planets in other solar systems have been found, but so far none of them appear to be like Earth. In order for a planet to be able support plant and animal life, the planet would need to be close to a sun. However, if it were too close, the planet would be too hot. If the planet were too far away from the sun, it would be too cold. Also, the sun would have to be large enough to produce enough energy for plants and animals to use.

Scientists continue to search for life on other planets. So far, Earth is the only planet that we know supports plant and animal life.

8.6 Summary

- Our nearest neighboring star is Proxima Centauri.

- The brightest star is Sirius.

- The largest star is VY Canis Majoris (VY CMa).

- So far, Earth is the only planet we know of that supports plant and animal life.

8.7 Some Things to Think About

- How many stars do you think exist and how many of those stars might have an Earth-like planet?

- Which is your favorite star? Why?

  Our Sun
  Proxima Centauri
  Alpha Centauri A
  Alpha Centaure B

- What is the brightest star you have seen in the night sky?

- What is the biggest star you have seen in the night sky? What makes it the biggest?

- How many Earth-like planets do you think might exist in the universe? Why?

Chapter 9 Galaxies

9.1   Introduction   72

9.2   Galaxies Are Like
       Cities in Space   72

9.3   How Many Galaxies?   73

9.4   What Is a Galaxy Made Of?   74

9.5   Other Stuff About Galaxies   75

9.6   Summary   76

9.7 Some Things to Think About 76

HOAG'S OBJECT

YOU ARE HERE

GALAXY CENTER

9.1 Introduction

You probably know that Earth is the third planet from the Sun and that there are seven other planets that orbit the Sun. You know that Earth "lives" in a "neighborhood" called a solar system. But where does our solar system "live"?

Solar systems, like ours, exist in a larger collection of stars, planets, and moons called a galaxy. Galaxies are like big cities, with several hundred or even thousands of solar "neighborhoods."

9.2 Galaxies Are Like Cities in Space

On a clear night, far away from city lights, you can often see a band of stars and light across the night sky. These stars are in the Milky Way Galaxy, the galaxy where we live. The Milky Way Galaxy is like a big city of stars, planets, dust, and other objects.

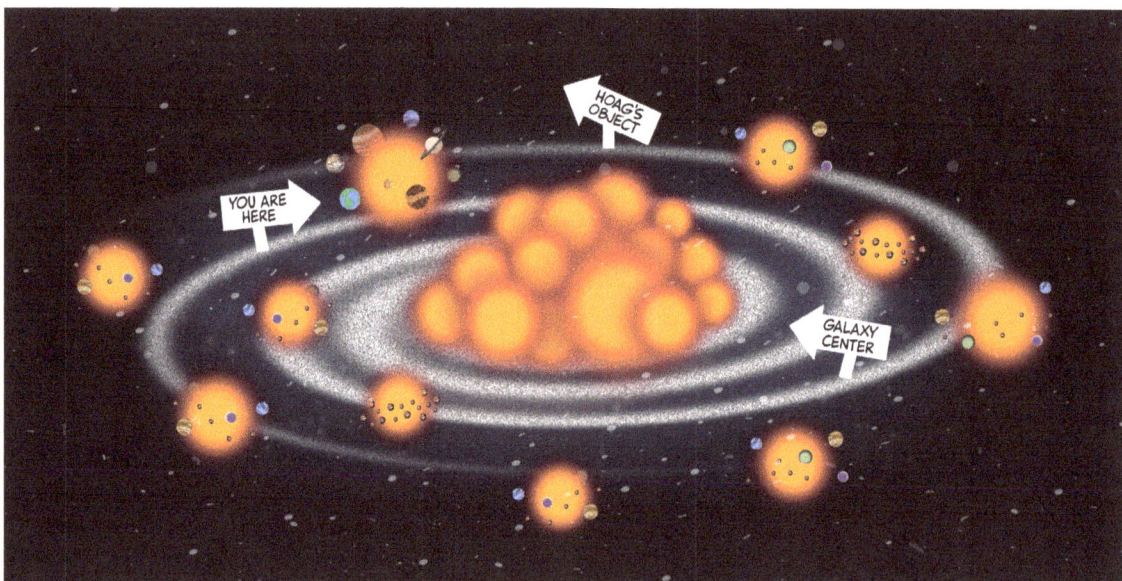

In a city, there are different neighborhoods, parks, shopping areas, and roads. All of these are grouped together to make a city. A city is organized in a certain way and has a certain shape that depends on how the neighborhoods, parks, shops and roads are put together.

The same is true of a galaxy. Like a big city, a galaxy holds all of the stars, planets, and other objects together in a particular shape. We will learn about the different shapes of galaxies in Chapter 11.

9.3 How Many Galaxies?

Scientists don't have a way to actually count the number of galaxies in the universe, but some astronomers estimate that there may be at least 170 billion galaxies. Other astronomers think there may be a trillion galaxies or even more! That's a lot of galaxies!! As our space telescopes become more and more powerful, astronomers think they will be able to see more and more galaxies.

It's hard to imagine just how many galaxies 170 billion would be. If you pretend that one galaxy is the size of a marble and if you had 170 billion

marbles and lined them up end-to-end, you would need to wrap them around the Earth almost 30 times! That's a LOT of marbles!

9.4 What Is a Galaxy Made Of?

Galaxies are made of gases, stars, planets, dust, and other objects. Some scientists believe that most galaxies have a black hole at the center with the stars and other objects revolving around the black hole. This theory holds that within galaxies new stars are formed from clouds of gas and dust and that most galaxies have enough gas and dust to form billions of new stars.

Galaxies are thought to begin as small clumps of stars and then grow as new stars are made. Scientists also believe that stars can get old and can explode at the end of their life.

9.5 Other Stuff About Galaxies

Astronomers have discovered some very small galaxies that contain only about 1000 stars. They have also discovered gigantic galaxies that are 50 times the size of the Milky Way! The Milky Way is considered to be an average size galaxy.

Another interesting fact is that everything in space is in motion. For most galaxies, the objects within the galaxy revolve around the center of the galaxy. The galaxies themselves are moving through space, and scientists believe that sometimes galaxies will run into each other and join to form a new galaxy.

9.6 Summary

● A galaxy is like a big city made of lots of stars, planets, dust, and other objects.

● We live in the Milky Way Galaxy.

● There are many billions of galaxies in the universe.

● Galaxies come in many different sizes, from very small to gigantic, with the Milky Way being an average size galaxy.

9.7 Some Things to Think About

● Do you think ancient astronomers were able to tell that Earth is in a galaxy? Why or why not?

● Do you think galaxies can come in different shapes or do they all have to be the same? Why?

● How do you think powerful telescopes have helped astronomers make new discoveries about galaxies and about the number of galaxies that may exist?

● What do you think makes galaxies grow?

● What do you think would make galaxies run into each other?

Chapter 10 Our Galaxy The Milky Way

10.1   Introduction   78

10.2   Our Galaxy   78

10.3   Where Are We?   80

10:4   Earth Moves   81

10.5   Summary   82

10.6   Some Things
       to Think About   82

PERSEUS ARM

SAGITTARIUS ARM

NORMA ARM

SCUTUM–CENTAURUS ARM

10.1 Introduction

In this chapter we will take a closer look at our galaxy, the Milky Way Galaxy. Scientists estimate that there are many billions of stars in the Milky Way Galaxy and that many of these stars may have planets orbiting them.

10.2 Our Galaxy

Think about the city you live in or one you have visited. Can you see the whole city from your house? Is it easy to tell the shape of your city from where you live?

No! Because you are one small person in a big city, you can't tell what your city looks like from your house. You would need to see your city from a different place, like an airplane or spaceship, to see what shape it has.

In the same way, it is difficult for astronomers to see the Milky Way Galaxy. No one has taken a picture of the Milky Way Galaxy because it is too big and we can't fly far enough away to see the whole galaxy. However, astronomers can guess what the Milky Way Galaxy looks like by observing other galaxies.

Is our galaxy round or flat? Is our galaxy large or small? Does our galaxy have a fixed center, like an orange, or does it move like Jell-O?

Even though we've never seen our galaxy from the outside, modern astronomers think that the Milky Way is shaped like a pinwheel. Just like a pinwheel, our galaxy has spiraling arms and a bulge in the center.

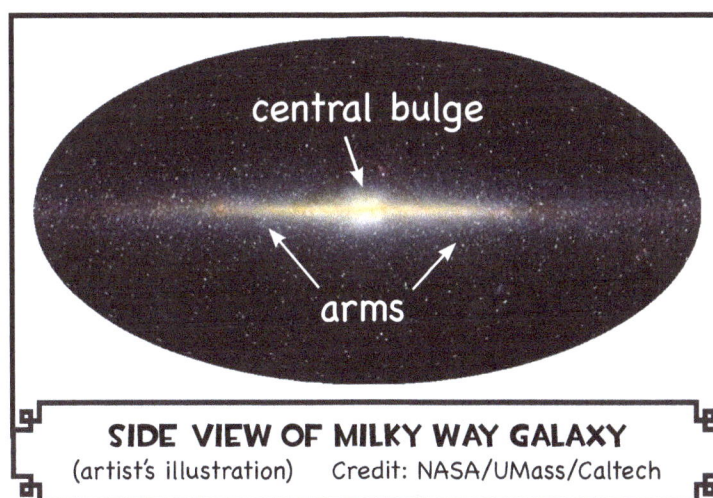

SIDE VIEW OF MILKY WAY GALAXY
(artist's illustration)   Credit: NASA/UMass/Caltech

This central bulge is a dense ball of stars. The arms of our galaxy are flatter at the edges than the center. Most of the stars in our galaxy are in the center, with fewer stars on the edges.

The Milky Way has two major arms, which are called the Scutum-Centaurus Arm and the Perseus Arm, and two minor arms, called the Norma Arm and the Sagittarius Arm. These arms spread out from the center, creating a spiral galaxy that looks like a pinwheel.

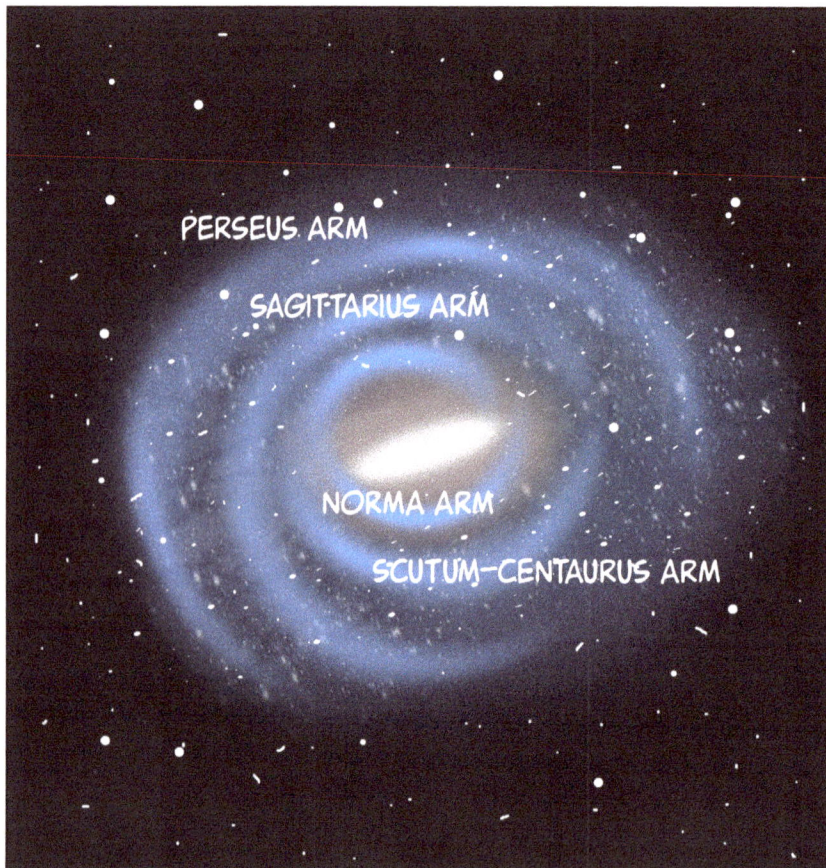

10.3 Where Are We?

Our Sun and solar system are located on a partial arm called the Orion Arm. The Orion Arm is between the Sagittarius and Perseus arms. Scientists think that our solar system may be about halfway between the center and the outer edge of the Milky Way Galaxy, but this is still uncertain.

We happen to live at just the right place in our galaxy. If our solar system were too far from the center of the galaxy, a planet like Earth might not be able to form. If our solar system were too close to the center, there might be too many stars creating too much radiation and gravity for life to form. As it turns out, we live in the right place in our galaxy for life to exist—not too close to the center and not too far away!

10.4 Earth Moves

The Milky Way Galaxy, like other galaxies, is in motion. It is thought that the objects in the Milky Way Galaxy revolve around a black hole at the center and that the entire Milky Way Galaxy moves through space. We can see that Earth, too, is constantly in motion. The Earth spins on its axis, revolves around the Sun, travels with our solar system around the center of the Milky Way Galaxy, and moves through space with the entire galaxy.

10.5 Summary

● Astronomers think the Milky Way Galaxy is shaped like a pinwheel.

● The shape of the Milky Way Galaxy is called a spiral galaxy.

● Our Earth is in the right spot in our galaxy for life to exist.

● Earth moves through space in several ways—spinning on its axis, revolving around the Sun, traveling with our solar system around the center of the Milky Way Galaxy, and moving through space with the entire Milky Way Galaxy.

10.6 Some Things to Think About

● If there are billions of stars in the Milky Way Galaxy and if many of these stars have planets orbiting them, do you think these planets would be similar to the planets in our solar system? Why or why not?

Do you think some of these other planets might be totally different kinds of planets? If so, what do you think different kinds of planets might be like?

⬤ Do you think there might be solar systems other than ours that are in the right place in the Milky Way Galaxy for life to exist? Why or why not?

⬤ How do you think you can tell that Earth is moving?

Chapter 11 Beyond Our Galaxy

11.1   Introduction   85

11.2   More Sprial Galaxies   85

11.3   Other Types of Galaxies   86

11.4   The Local Group of Galaxies   88

11.5   Summary   89

11.6   Some Things to Think About   89

11.1 Introduction

In Chapter 10 we looked at our galaxy, the Milky Way Galaxy. We saw that our galaxy is called a spiral galaxy and that it has a central bulge and has arms that extend outward from the center like a pinwheel.

In this chapter we will take a look at other types of galaxies. From Earth, astronomers have been able to view thousands of different galaxies. Some of these galaxies look like ours, with a central bulge and spiral arms. But some galaxies look very different from our galaxy and have unusual features.

11.2 More Spiral Galaxies

There are many galaxies like ours. Spiral galaxies are fairly common in the universe. However, even spiral galaxies look different from one another.

In some spiral galaxies the central bulge is very large. This type of galaxy is called an Sa galaxy. Other spiral galaxies have a central bulge that is smaller. This type of galaxy is called an Sc galaxy. What the arms look like is also taken into consideration when classifying spiral galaxies.

Some spiral galaxies have a bar-shaped cluster of stars in the center. This type of spiral galaxy is called a barred spiral galaxy, or an SB galaxy. Many astronomers think the Milky Way Galaxy might be a barred spiral galaxy rather than a regular spiral galaxy.

11.3 Other Types of Galaxies

Astronomers have seen other types of galaxies that are different from spiral and barred spiral galaxies.

An elliptical galaxy is a type of galaxy that can look like one huge star but is really a group of tightly packed stars. Elliptical galaxies can vary greatly in size. Some elliptical galaxies are small and are called dwarf elliptical galaxies.

There are other elliptical galaxies that are very large and are made of trillions of stars. Elliptical galaxies are round or elliptical in shape and don't have any special features.

Irregular galaxies don't look like spiral galaxies or elliptical galaxies. They don't really fit into any other category of galaxies and have a variety of odd features.

Some irregular galaxies have a large bulge of stars off to one side with a ring of stars surrounding it. Other irregular galaxies are dumbbell or butterfly shaped.

It's hard to know how some of these galaxies got their shape. Some astronomers think it might be possible that some of these irregular galaxies have such odd shapes because they are galaxies that have bumped into each other. But other astronomers think that the irregular shapes developed as the galaxies formed. Learning how galaxies form is an exciting area of study in astronomy.

11.4 The Local Group of Galaxies

Just as planets exist together around a star to form a solar system and solar systems exist together to form galaxies, astronomers have discovered that galaxies exist together to form large groups. The Milky Way Galaxy is actually part of a large group of galaxies that are close together. Astronomers call this group of galaxies the Local Group. It's uncertain how many galaxies are in the Local Group because astronomers keep discovering new galaxies that are close to ours. Some astronomers now think there are over 50 galaxies in the Local Group.

The Milky Way is one of the three largest galaxies in the Local Group. Andromeda, our nearest neighboring galaxy, is the biggest galaxy in the Local Group. And Triangulum is the third largest galaxy in the Local Group, with the Milky Way being the second largest.

11.5 Summary

● The three types of galaxies are spiral, elliptical, and irregular.

● Spiral galaxies can have large or small central bulges and can have a bar-shaped cluster of stars in the center.

● An elliptical galaxy can look like one huge star but can contain billions or trillions of stars.

● Galaxies exist together to form large groups.

11.6 Some Things to Think About

● After learning about the Milky Way Galaxy, what feature of a spiral galaxy do you think is the most interesting?

Billions of stars
Central bulge of stars
Spiral arms
Black hole
Shape of the galaxy
Motion of the galaxy
Solar systems
Something else

● Why do you think it helps astronomers to group galaxies by their type?

● Do you think galaxies can be any shape? Why or why not?

Do you think galaxies can change from one shape to another? Why or why not?

Do you think there are rules that galaxies have to follow when they form in a particular shape? Why or why not?

● Do you think all galaxies are arranged in groups or are some by themselves? Why?

Chapter 12 Other Stuff in Space

12.1   Introduction   92

12.2   Comets and Asteroids   93

12.3   Exploding Stars   94

12.4   Collapsed Stars   95

12.5   Nebulae   95

12.6   Summary   97

12.7   Some Things
       to Think About   97

LET'S GO CHECK OUT THE CRAB NEBULA

OKAY!

12.1 Introduction

In Chapter 11 we looked at different types of galaxies. We saw that some galaxies are like ours, with a central bulge and spiral arms. We also saw that some galaxies are irregular or elliptical.

What other things exist in the universe besides stars, planets, and galaxies?

12.2 Comets and Asteroids

Comets and asteroids are found throughout the universe. Comets are large chunks of rock and ice that fly though space at great speeds. When a comet gets near a sun, the heat will make the ice in the comet begin to vaporize, creating a beautiful tail. When ice vaporizes, it changes to a gas without becoming liquid water first. You can notice the results of vaporization if you leave ice cubes in the freezer for a long time. They get smaller as the ice vaporizes.

Asteroids are made of rock and have irregular shapes. The Asteroid Belt is a ring of asteroids that travel around the Sun between the orbits of Mars and Jupiter. Many other asteroids exist outside the Asteroid Belt. Like comets, asteroids move through space at very high speeds. Most asteroids are small, and sometimes they collide with one another! As a result asteroids are often covered with small craters that are caused by these impacts.

Once in a while a comet or an asteroid will come close enough to Earth to hit it. If an asteroid hits Earth, it is called a meteorite. Although asteroids occasionally hit Earth, most of the time they burn up in our atmosphere before they reach the ground. If you have ever seen a shooting star, it is really an asteroid burning up as it flies through Earth's atmosphere.

12.3 Exploding Stars

Stars do not stay the same size or generate the same energy forever. Stars actually have a birth and a death. When a star is born, it is able to generate light and heat energy for a very long period of time. However, at some point it runs out of energy and dies. When the star begins to run out of energy, it gets very large, burning brighter and brighter. Astronomers call this type of star a red giant. Once the red giant star uses up all its energy, it shrinks into a small white dwarf star.

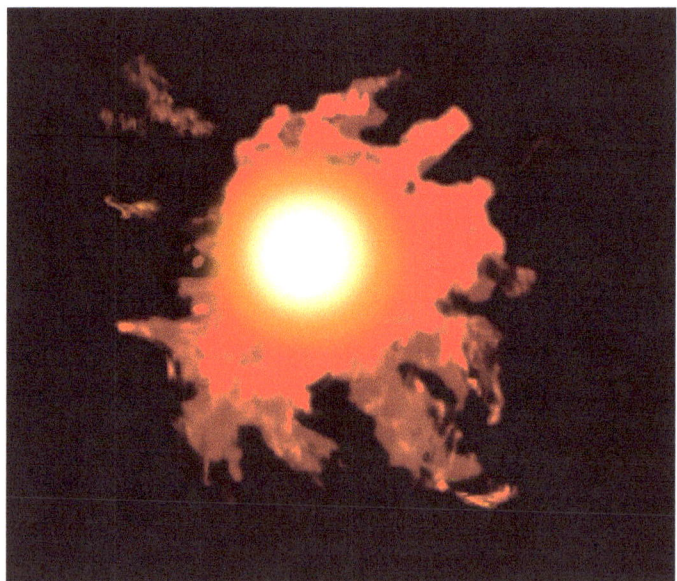

Sometimes stars actually explode. A supernova is a star that is exploding. When a supernova star explodes, it becomes very large and bright, expanding many millions of miles into the surrounding area.

12.4 Collapsed Stars

What happens to the exploding star once it has finished exploding? Where does it go? Does it turn into nothing, or does it become something else?

Many astronomers think that after a big star explodes, it may collapse and form a black hole. A black hole is an odd feature in the universe. It is difficult to see a black hole because it doesn't allow any radio waves or light waves to bounce back from it. Because of this, it just looks like there is a dark, black hole in the middle of space.

12.5 Nebulae

By using the Hubble Space Telescope, astronomers can explore the universe in ways never before possible. The Hubble Space Telescope has given us some very beautiful pictures of stars, planets, asteroids, and galaxies.

Some of the most beautiful images captured by the Hubble Space Telescope are nebulae. Nebulae are clouds of gas, dust, and particles. The gas, dust, and particles swirl in space to create amazing celestial sculptures and cosmic art. Today several thousand nebulae have been photographed with the Hubble Space Telescope.

However, we have only just begun imaging, understanding, and discovering the stars, planets, and other objects that exist in space. Future generations of astronomers have a whole universe to discover and explore!

12.6 Summary

- Comets are objects in space that are made of rock and ice.

- Asteroids are objects in space that are made of rock and have irregular shapes. If an asteroid makes contact with Earth, it is called a meteorite.

- A star that explodes is called a supernova.

- Black holes are thought to be collapsed supernova stars.

- Nebulae are clouds of gas, dust, and particles.

12.7 Some Things to Think About

- Do you think we have already discovered all the different types of objects that exist in the universe? Why or why not?

- What do you think happens to a comet as its water vaporizes? Why?

  What do you think happens when a meteorite hits Earth? Why?

  Do you think the size of the asteroid the meteorite came from makes a difference? Why?

- Do you think a million miles is a big distance compared to the whole universe? Why or why not?

● If astronomers can't see a black hole, how do you think they know it is there?

● What stuff in space would you most like to find out more about? Why?

Galaxies

Nebulae

Comets

Asteroids

Supernovae

Black holes

Clouds of gas and dust

Glossary–Index

Alpha Centauri A (AL-fuh sen-TAW-ree AY) • a star close to Earth that is in the Alpha Centauri star system, 66

Alpha Centauri B (AL-fuh sen-TAW-ree BEE) • a star close to Earth that is in the Alpha Centauri star system, 66

Andromeda (an-DRAH-meh-duh) • the biggest galaxy in the Local Group, 88

Archer (AR-chuhr) • a constellation in the Southern Hemisphere, 45

Aristarchus of Samos (aah-ruh-STAHR-kuss of SAY-mahs) • [310-230 BCE } Greek astronomer and mathematician; said that the Earth travels around the Sun and also revolves around its own axis, 4

asteroid (AS-tuh-roid) • a small, irregularly shaped rocky object in space, 93-94

Asteroid (AS-tuh-roid) **Belt** • a ring of asteroids that travel around the Sun between the orbits of Mars and Jupiter, 93

astronomy (uh-STRAH-nuh-mee) • the study of the cosmos, 2

axis (ACK-sis) [plural, **axes** (ACK-seez)] • an imaginary line that goes through the center of an object, 5, 20-21, 46, 52, 81

barred spiral galaxy • see galaxy, barred spiral

basic tool • a simple tool, 11

Big Dipper • a constellation in the Northern Hemisphere, 45, 47, 48

black hole • an area in space that doesn't allow any radio waves or light waves to bounce back from it, 74, 81, 95

Brahe, Tycho (BRAY-hee, TEE-koh) • [1546-1601 CE] Danish astronomer; created a map of the stars, 6-7, 8

Cassiopeia (kaa-see-uh-PEE-uh) • a constellation in the Northern Hemisphere, 45

central bulge (SEHN-truhl BUHLJ) • a dense ball of stars at the center of a spiral galaxy, 79, 85, 86

chemical reaction (KEH-muh-kul ree-ACK-shuhn)• occurs when the electrons in atoms join or break apart and the neutrons and protons stay the same, 33

comet (KAH-muht) • a large chunk of rock and ice that flies though space, 93, 94

computer (kum-PYOO-ter) • an electronic device that can store, retrieve, and process data, 11

constellation (kon-stuh-LAY-shuhn) • a group of stars that forms a shape or image in the sky, 45-52

Copernicus, Nicolaus (koh-PURR-ni-kuss, NI-koh-lahs) • (1473-1543 CE) Polish astronomer; used mathematics to show that the Earth moves around the Sun, 5-6, 7, 8, 11-12

cosmos (KAHZ-mohs) • the Earth and everything that extends beyond the Earth, 2

crater (CRAY-ter) • a bowl-shaped depression around the opening of a volcano; a hole in the ground from a meteorite impact, 30, 39, 93

crescent (CRE-sent) • the shape of the Moon when less than half of it is visible, 29

dwarf elliptical galaxy • see galaxy, dwarf elliptical

dwarf planet • see planet, dwarf

Earth • the planet we live on, 4-7, 19-25, 28-33, 37, 39, 46, 47, 52, 57, 59-61, 81

eclipse, lunar (i-CLIPS, LOO-ner) • a darkening of the Moon as the Moon passes behind the Earth and the Earth's shadow falls on the Moon, 25

eclipse, solar (i-CLIPS, SOH-ler) • a darkening of the Sun when the Moon passes between the Earth and the Sun and the Sun's rays are blocked from reaching Earth, 25

elliptical galaxy • see galaxy, elliptical

eyepiece (EYE-peess) • the lens at one end of a telescope that the eye looks through, 13

full moon • the phase of the Moon during which the Moon looks like a full, round circle, 29

galaxy (GAH-lek-see) • a large group of stars, gas, dust, planets, and other objects in space, 15, 72-75, 78-81, 85-88

galaxy (GAH-lek-see), **barred spiral** • a spiral galaxy that has a bar-shaped cluster of stars in the center, 86

galaxy, dwarf elliptical (GAH-lek-see, DWAWRF i-LIP-ti-kuhl) • a small elliptical galaxy, 86

galaxy, elliptical (GAH-lek-see) • a group of tightly packed stars that looks like one huge star, 86

galaxy, irregular (GAH-lek-see i-REH-gyuh-ler) • a galaxy that doesn't fit in any other category, 87

galaxy, Sa (GAH-lek-see, ESS AY) • a spiral galaxy with a very large central bulge, 86

galaxy, SB (GAH-lek-see, ESS BEE) • a barred spiral galaxy, 86

galaxy, Sc (GAH-lek-see, ESS SEE) • a spiral galaxy with a small central bulge, 86

galaxy, spiral (GAH-lek-see, SPY-rul) • a galaxy that is shaped like a pinwheel, 80, 85-86

Galileo [Galileo Galilei (ga-luh-LAY-oh gal-uh-LAY)] • [1564-1642] Italian scientist; used mathematics and physics to prove we live in a Sun-centered solar system, 7-8, 12

gravity (GRA-vuh-tee) • the force that holds everything to Earth's surface, 20, 23, 37, 69, 81

Great Bear • a constellation in the Northern Hemisphere, 48

Great Square • a constellation in the Southern Hemisphere, 49, 50

Gregorian telescope • see telescope, Gregorian

Hubble Space Telescope (HUH-buhl SPAYSS TEL-uh-scope) • a telescope that is in orbit around the Earth outside the atmosphere, 15, 95-96

IAU (International Astronomical Union) • see International Astronomical Union

illuminate (i-LOO-muh-nayt) • to light up, 67

inner solar system • see solar system, inner

instrument (IN-struh-ment) • a tool that is used to do careful and precise work, such as taking measurements, 13

International Astronomical Union (IAU) (in-ter-NASH-uh-null as-trah-NAH-mi-kull YOON-yun) • a group of professional astronomers from around the world that determines how celestial bodies should be defined and classified, 41-42

irregular galaxy • see galaxy, irregular

Jovian (JOH-vee-un) • Jupiter-like, 38, 40

Jupiter (JOO-puh-ter) • a large, gaseous planet in the Earth's solar system, 37, 40, 57, 58, 60, 93

landmark • an object that marks a location and can help with navigation, 51

lava flow (LAH-vuh floh) • molten or hardened lava around the mouth of a volcano on a moon or planet, 30

lens (lenz) • a transparent glass or plastic object that has one to two curved sides and is used in telescopes to make objects appear larger, 6, 12, 13-15

Lippershey, Hans (LI-per-shee, HAHNZ) • (1570-1619 CE) a Dutch lens maker; invented the spyglass, 12

Local Group • the large group of galaxies that includes the Milky Way Galaxy, 88

lunar eclipse • see eclipse, lunar

magnify (MAG-nuh-fye) • to make something appear larger, 13

Mars • a terrestrial planet in our solar system, the fourth from the Sun, 37, 38, 39, 57, 59-60, 93

Mercury (MER-kyuh-ree) • a terrestrial planet in our solar system; the closest planet to the Sun, 37, 39, 56, 57, 59-60

meteorite (MEE-tee-uh-rite) • an asteroid that hits the Earth, 94

Milky Way Galaxy (GAH-lek-see) • the galaxy where we live, 72, 75, 78-81, 85, 86, 88

moon • a natural object that travels around a planet, 3, 4, 23, 25, 28-30, 37, 72

navigation (na-vuh-GAY-shuhn) • making one's way from one location to another, 47, 50-52

nebula (NE-byuh-luh) [plural, **nebulae** (NE-byuh-lee)] • a cloud of gas, dust, and particles in space, 95-96

Neptune (NEP-toon) • a Jovian planet in the Earth's solar system, eighth from the Sun, 37, 38, 40, 58, 59-60

Newtonian telescope • see telescope, Newtonian

Norma Arm • one of the minor spiral arms of the Milky Way Galaxy, 80

North Star • the star that is above the northernmost point of the Earth, can be used as a navigational marker; also called Polaris, 47

nuclear reaction (NOO-klee-er ree-ACK-shuhn) • occurs when the protons and neutrons of an atom are rearranged, creating energy and changing the structure of the atom, 33

objective lens (ahb-JEK-tiv LENZ) • the lens of a telescope that collects light so an object can be viewed, 13, 15

orbit (AWR-bit) • the curved path of a celestial body as it travels around another celestial body, 15, 24, 25, 29, 37, 58-60, 69, 72

Orion (uh-RYE-un) **Arm** • the small, partial spiral arm where Earth is located in the Milky Way Galaxy, 80

Orion (uh-RYE-un) **the Hunter** • a constellation in the Northern Hemisphere, 48

outer solar system • see solar system, outer

Pegasus (PE-guh-sus) • a constellation in the Southern Hemisphere, 50

Perseus Arm (PER-see-us ARM) • one of the major spiral arms of the Milky Way Galaxy, 80

planet (PLA-net) • a large spherical celestial body that moves around (orbits) a sun and is large enough to have its own gravity, 20, 35-41, 55-61, 69-70, 72, 78, 81

planet, dwarf • a celestial body that does not have all the features needed to be classified as a planet; also called a plutoid, 41

planet, Jovian (JOH-vee-un) • a large planet that is similar to Jupiter; also called a gas giant; Jupiter, Saturn, Uranus, and Neptune, 38, 40

planet, terrestrial (PLA-net, tuh-RES-tree-ul) • a planet that is Earth-like; made of rocky materials and closer to our Sun; Mercury, Venus, Earth and Mars, 38-39, 55

planetary (PLA-nuh-te-ree) **year** • the number of days it takes a planet to orbit the Sun, 59-60

Pluto (PLOO-toe) • once considered the 9th planet, is now classified as a dwarf planet or plutoid, 41-42

plutoid (PLOO-toid) • a celestial body that does not have all the features needed to be classified as a planet; also called a dwarf planet, 41

Polaris (puh-LER-us) • the star that is above the northernmost point of the Earth, can be used as a navigational marker; also called the North Star, 47, 52

prove (proov) • to establish the truth of something, 7, 8

Proxima Centauri (PROCK-suh-muh sen-TAW-ree) • the star that is nearest to our solar system; part of the Alpha Centauri system, 66, 67

red giant star • see star, red giant

rotate (ROE-tate) • to move around an axis or a center, 20

Sa galaxy • see galaxy, Sa

Sagittarius (sa-juh-TERR-ee-us) **Arm** • one of the minor spiral arms of the Milky Way Galaxy, 80

satellite (SA-tuh-lite) • a celestial body that orbits a larger celestial body; a machine that is put into orbit around the Earth or another celestial body, 11

Saturn (SA-turn) • a Jovian planet in the Earth's solar system, sixth from the Sun; has rings, 37, 38, 40, 58, 60, 68

SB galaxy • see galaxy, SB

Sc galaxy • see galaxy, Sc

Scutum-Centaurus Arm (SCOO-tum sen-TAW-rus ARM) • one of the major spiral arms of the Milky Way Galaxy, 80

sighting tube (SIE-ting TOOB) • a hollow tube that does not have lenses and is used to look at stars, 6-7

Sirius (SIR-ee-us) • the brightest star in the sky, 67

solar eclipse • see eclipse, solar

solar system (SOH-ler SIS-tem) • a single sun and the group of celestial bodies that orbit it, 7, 15, 36-42, 55-61, 65, 69, 80, 81

solar system (SOH-ler SIS-tem)**, inner** • the four planets closest to the Sun—Mercury, Venus, Earth, and Mars, 57

solar system (SOH-ler SIS-tem)**, outer** • the four planets farthest from the Sun—Jupiter, Saturn, Uranus, and Neptune, 57-58

Southern Cross • a constellation in the Southern Hemisphere that can be used for navigation, 52

spaceship (SPAYS-ship) • a vehicle that can travel beyond Earth; 11

spherical (SFIR-i-kul) • ball-shaped, 20

spiral (SPY-rul) **arms** • in a galaxy, groups of stars that extend from the galactic bulge in a pinwheel-like form, 79, 85

spiral galaxy • see galaxy, spiral

spyglass (SPY-glass) • a tube with a lens at each end that is used to make stars look bigger, 12

star • an object in space that generates its own light and heat energy, 30-33, 45-52, 65-70, 72, 74-75, 78, 79, 86, 87, 94-95

star, red giant • a star that has begun to run out of energy, has gotten larger, and is burning brighter and brighter, 94

star, white dwarf • a small star that results when a red giant star has used up all its energy, 94

sun • a star, 30-33

supernova (soo-per-NOH-vuh) [plural **supernovae** (soo-per-NOH-vee)] • an exploding star, 94

telescope (TE-luh-scope) • a tool or instrument that helps astronomers see far into the distance; makes faraway objects look larger, 11-15, 66, 73, 95-96

telescope, advanced (TE-luh-scope, ad-VANST) • a more complicated telescope that is able to see objects farther away in space, 15

telescope, Gregorian (TE-luh-scope, gri-GAWR-ee-un) • a telescope that has mirrors on the inside of the tube, 14-15

telescope, Newtonian (TE-luh-scope, new-TOH-nee-un) • a telescope that has mirrors on the inside of the tube, 14-15

terrestrial (te-RES-tree-ul) • Earth-like, 38, 39, 55

terrestrial planet • see planet, terrestrial

tilt • slant, 21

Triangulum (trie-AN-gyuh-lum) • the third largest galaxy in the Local Group, 88

Uranus (YOOR-uh-nus) • a Jovian planet, seventh from the Sun in the Earth's solar system, 37, 38, 40, 58, 60

vaporize (VAY-puh-rize) • to change from a solid to a gas without becoming a liquid first, 93

Venus (VEE-nus) • a terrestrial planet, the second from the Sun in the Earth's solar system, 37, 38, 39, 56-57, 59-60

VY Canis Majoris (VEE WYE KAY-nus muh-JAWR-us) **(VY CMa)** • the biggest star we can see from Earth, 68

Whale • a constellation in the Southern Hemisphere, 45, 49

white dwarf star • see star, white dwarf

More REAL SCIENCE-4-KIDS Books
by Rebecca W. Keller, PhD

Building Blocks Series
yearlong study program — each Student Textbook has accompanying Laboratory Notebook, Teacher's Manual, Lesson Plan, Study Notebook, Quizzes, and Graphics Package

Exploring the Building Blocks of Science Book K (Activity Book)
Exploring the Building Blocks of Science Book 1
Exploring the Building Blocks of Science Book 2
Exploring the Building Blocks of Science Book 3
Exploring the Building Blocks of Science Book 4
Exploring the Building Blocks of Science Book 5
Exploring the Building Blocks of Science Book 6
Exploring the Building Blocks of Science Book 7
Exploring the Building Blocks of Science Book 8

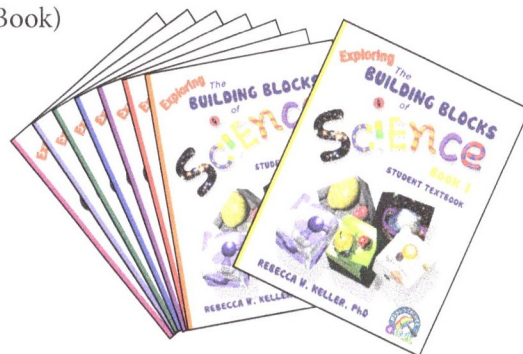

Focus Series
unit study program — each title has a Student Textbook with accompanying Laboratory Notebook, Teacher's Manual, Lesson Plan, Study Notebook, Quizzes, and Graphics Package

Focus On Elementary Chemistry
Focus On Elementary Biology
Focus On Elementary Physics
Focus On Elementary Geology
Focus On Elementary Astronomy

Focus On Middle School Chemistry
Focus On Middle School Biology
Focus On Middle School Physics
Focus On Middle School Geology
Focus On Middle School Astronomy

Focus On High School Chemistry

Super Simple Science Experiments

21 Super Simple Chemistry Experiments
21 Super Simple Biology Experiments
21 Super Simple Physics Experiments
21 Super Simple Geology Experiments
21 Super Simple Astronomy Experiments
101 Super Simple Science Experiments

Note: A few titles may still be in production.

Gravitas Publications Inc.
www.gravitaspublications.com
www.realscience4kids.com

GRAVITAS
PUBLICATIONS

www.ingramcontent.com/pod-product-compliance
Lightning Source LLC
Chambersburg PA
CBHW050240220326
41598CB00047B/7464